THE NEW LUNAR SOCIETY

THE NEW LUNAR SOCIETY

AN ENLIGHTENMENT GUIDE TO THE NEXT INDUSTRIAL REVOLUTION

DAVID A. MINDELL

The MIT Press
Cambridge, Massachusetts
London, England

The MIT Press would like to thank the anonymous peer reviewers who provided comments on drafts of this book. The generous work of academic experts is essential for establishing the authority and quality of our publications. We acknowledge with gratitude the contributions of these otherwise uncredited readers.

This book was set in Adobe Garamond Pro by New Best-set Typesetters Ltd. Printed and bound in the United States of America.

Library of Congress Cataloging-in-Publication Data

Names: Mindell, David A., author.
Title: The new lunar society : an enlightenment guide to the next
  industrial revolution / David A. Mindell.
Description: Cambridge, Massachusetts : The MIT Press, [2025] |
  Includes bibliographical references and index.
Identifiers: LCCN 2024024907 (print) | LCCN 2024024908 (ebook) |
  ISBN 9780262049528 (hardcover) | ISBN 9780262381888 (epub) |
  ISBN 9780262381895 (pdf)
Subjects: LCSH: Industrialization—United States. | Industries—
  United States. | Telecommuting—United States.
Classification: LCC HD2329 .M56 2025 (print) | LCC HD2329 (ebook)
  | DDC 330.9/034—dc23/eng/20241016
LC record available at https://lccn.loc.gov/2024024907
LC ebook record available at https://lccn.loc.gov/2024024908

10 9 8 7 6 5 4 3 2 1

For the new industrialists
and their lunar societies,
improving the world

CONTENTS

1	WATT'S HEROIC INVENTION	1
2	ON THE CUSP OF INDUSTRIAL TRANSFORMATION	5
3	INDUSTRIAL HABITS OF MIND	7
4	BUILDING THINGS, NOT CLIMBING LADDERS	9
5	ON COUNTING REVOLUTIONS	11
6	A STEAMPUNK FAIRY TALE THAT ACTUALLY HAPPENED	15
7	SHADOWS OF INDUSTRIAL ENLIGHTENMENT	19
8	BIRMINGHAM AWAKE WITH TOYS	23
9	DESIGNED IN CALIFORNIA . . . MADE WHERE?	27
10	FABRICATING THE PIN FACTORY	31
11	THE LIFEBLOOD OF INDUSTRIAL SOCIETIES	37
12	"MASTER OF EVERY METALLIC ART"	39
13	INDUSTRY	47
14	THE INDUSTRIOUS REVOLUTION	51
15	THE GREAT TOILET PAPER CRISIS	55
16	SUPPLY CHAINS ARE US	57
17	JEFFERSON'S ENLIGHTENMENT TEACHER	61
18	CONQUERING WITH POTTERY	65

19   "TURNING THE MILL BY FIRE"   69

20   THE WORK OF THE FUTURE   73

21   JAMES WATT, FRAIL CRAFTSMAN   81

22   AUTOMATION AND WORK   91

23   "TO SETTLE A MANUFACTORY"   95

24   HOW YOU GET TO WORK   101

25   DEATH FORMS THE LUNAR SOCIETY   105

26   "STEAM MILL MAD"   113

27   "GET EXCITED ABOUT MAINTENANCE"   127

28   JEFFERSON'S LOST WORLD   131

29   CULTS OF NEWNESS AND THE CHALLENGE OF
      ADOPTION   135

30   REINDUSTRIALIZING AMERICA   137

31   "TO ASTONISH THE WORLD, ALL AT ONCE"   145

32   GENERATIONS OF INDUSTRIAL
      TRANSFORMATION   161

33   HUMAN MACHINES AT ETRURIA   163

34   A REPUBLIC OF INDUSTRY   171

35   FROM REVERE TO R&D   177

36   BEYOND BOULTON AND WATT: STEAM FOR
      MOBILITY   181

37   AN R&D SYSTEM ADRIFT   185

38   MANUFACTURING SOCIAL CHANGE   187

39   "NO PHILOSOPHERS!"   191

40   NEW INDUSTRIALISTS   195

41   LUNAR SOCIETIES TODAY   197

Contents

42   THE NEW INDUSTRIALISM   203

43   COLLABORATIVE HEROES   213

44   INDUSTRIAL FUTURES   217

Acknowledgments   219
Notes   221
Bibliography   243
Index   259

WATT'S HEROIC INVENTION

It is the founding mythology of the industrial world: James Watt, a young mechanic in Scotland, sets his mind to improving inefficient steam engines. For decades before him, the machines created by Thomas Newcomen, each the size of a building, had harnessed the power of steam and the weight of air to pump water out of mines. But they burned massive amounts of fuel, making them economical for only the largest, richest mines.

In a flash of insight on the green in his home town of Glasgow, Watt saw the source of waste: the engine heated up the massive cylinder every time steam rushed in to lift the piston, then immediately cooled the same cylinder by pumping in cold water to condense the steam and draw a vacuum. These heating and cooling cycles consumed fuel. Watt had the simple, energy-saving idea of creating a second cylinder to cool the steam. With this separate condenser, the main cylinder always stayed hot, and the engines consumed less fuel. Soon Watt's engines pumped out mines and drove factories, spawning the Industrial Revolution.

Economic historians recount this story as the earliest general purpose technology, akin to later electricity or computing, that influenced vast sectors of the economy and raised standards of living. A hundred years after his death, no less a figure than Watt's fellow Scot, steel magnate Andrew Carnegie, wrote one of the numerous heroic biographies of Watt.

Figure 1
Boulton and Watt lap engine from Soho Manufactory, including the cylinder and flyball governor. Photo by the author, London Science Museum.

In a British world used to honoring kings and admirals, Watt became a new modern person: practical and focused, mechanical and scientific. James Watt was the first engineer memorialized in Westminster Abbey. The word he invented underlies the very language of industry: the horsepower. Today's unit, the way to measure our way to a clean energy future, bears his name: the watt. He remains the archetype of the engineering hero mythology—a collective story about how technology drives history.

Watt deserves the accolades, and he is as interesting a figure as we will encounter in these pages. But the heroic genius mythology no longer serves well. The sole source for Watt's Glasgow green revelation is Watt's own memory, half a century later. It overemphasizes the lone inventor and the eureka moment, the simple invention from which all follows. It neglects the critical industrial substrates, the culture of ideas and machinery that shaped Watt and brought his ideas to fruition. It obscures the numerous other people who contributed to the engine to make it work.

Aspiring inventors today aim to emulate the Watt mythology, creating the pivotal invention that instantly changes the world. Too often those aspirations fail to grasp the richness and breadth essential for industrial change. At today's transformative moment, that failure narrows our vision. It's time for stories that are at once more faithful to the history and more useful for the future. If the Watt story is an origin myth for industrial revolution, updating it enables us to reimagine industrial life.

The Watt inventor story also pushes aside Matthew Boulton, Watt's entrepreneurial business partner. Upon closer inspection, Boulton emerges as the visionary of the pair. Commercially ambitious and accomplished at a young age, he manufactured small metal products at large scale long before he met Watt. When he came across the struggling Watt, he provided not only capital but know-how, access to fine tools, and skilled workmen.

Most important, Boulton recognized that Watt's invention could be applied to much more than pumping out mines, that it could transform manufacturing throughout the world. He also brought Watt into a circle of scientists and thinkers focused on bringing the ideals of the Enlightenment to the production of material goods: the Lunar Society. Watt's connections through this group, and his partnership with

Boulton, enabled him to complete his engine, to refine it with subsequent improvements, and produce and distribute engines to transform industry.

To use today's terminology, Watt's separate condenser was a *product innovation*. Boulton's ideas were *process innovations*. The Lunar Society was a social invention that connected product innovation to process innovation.

Industry in America today has too often lost this crucial link. This book is about the need to remarry product innovation to process innovation to imagine industrial futures, and about how to reimagine industrial futures to better serve human ends.

ON THE CUSP OF INDUSTRIAL TRANSFORMATION

Today again, industry stands on the threshold of a transformation. As an age of artificial intelligence seems to dawn, and an age of climate crisis is already here, we are poised at a moment that will shape not only commerce and industry but human subjectivity and behavior.

Yet at the same time, the nexus of energy, work, and technology that made industrial culture successful is failing the demands of the present and the future. Supply chain breakdowns, labor scarcity, and vulnerable manufacturing threaten economic security. Industry is responsible for most greenhouse gas emissions in the United States. Climate change compels us to reorient. This century requires a reinvention of industry in America.

Yet industries do not transform as quickly as software updates. It can take thirty years to change an entire system. To have any hope of carbon neutrality by 2050, or of systems resilient to global events, requires reinventing industry now.

For nearly three centuries, ideals inherited from the Industrial Enlightenment have driven industry. They have served us well: fed and clothed the masses, remade the continent, won wars, and offered prosperity to millions. But some of these ideals have also ravaged the land and the atmosphere, alienated workers, and shipped our vital skills overseas.

The disruptions of the 2020s make it clear that for America to decarbonize, it cannot deindustrialize. It must reindustrialize.

A host of new technologies are emerging from decades of experiments and slowly seeing adoption: robotics, artificial intelligence (AI), autonomy, electrification, and additive manufacturing among them. New ways of working, too: human-robotic collaboration, flexible work, remote and virtual presence. A new world must work more efficiently and more sustainably than anything that has come before. But how will it all fit together?

INDUSTRIAL HABITS OF MIND

Life in developed nations is industrial life. Societies and behaviors have arisen around fossil fuels and the power density they contain. This orientation goes much deeper than what kind of gas goes in our cars. Vaclav Smil calls them "the big four," the key commodities on which industrial society depends: ammonia, steel, plastics, and concrete. Each is indelibly shaped and created by fossil fuels, and any decarbonization strategy must address all four. Every time you see a concrete floor, or a plastic water bottle, or a steel skyscraper, think of the surging electrons that cooked the materials in a boiling furnace.

Fossil fuels shape the rhythms of our daily lives, our industrial habits of mind: suburban lifestyles; family weddings that convene relatives from great distances; college graduations and sporting events where thousands gather for a brief time and then disperse; vacations abroad; business trips; conferences and trade shows; instant ordering over the internet; fresh-cut grass and dust-free sidewalks; how nations defend themselves and make war. Nearly one half of 1 percent of all US energy use is military jet fuel, and the Department of Defense is the largest consumer of energy in the United States. Two percent of US energy is consumed in the extraction of oil and gas, and a further 1 percent of US energy goes toward transporting it to homes, cars, and businesses.

Anyone who depends on computers draws on global fossil-fueled semiconductor supply chains. Every mobile phone relies on an enormous global pyramid of materials and parts. Anyone who ever drives a car. We who eat strawberries in winter. One estimate shows that every tomato on your dinner plate consumes five tablespoons of crude oil to get there. A kilogram of chicken consumes half a bottle of wine's worth of oil. "Our food is partly made not just of oil, but also of coal, natural gas, electricity." You could say the same thing about your pharmaceuticals. These are today's habits of mind, ways that industry shapes our everyday behaviors.

The industrial systems that deliver these goods and experiences are of massive scale and are difficult to change quickly. Yet in 2020 the world shifted away from long commutes and office work, and it may never shift all the way back. Which other habits of mind could change more easily than we think?

BUILDING THINGS, NOT CLIMBING LADDERS

Corporate sales pitches of Industry 4.0 draw on the antiquated Watt mythology: individual inventions emerge from outside (startups in garages or geniuses in R&D centers), compelling society to resist or adopt them. Hype shows empty images with greenfield factories, isolation from broader systems, lacking human beings. Silicon Valley magnates brag of techno optimism, uninformed about industry or culture.

Utopian sales pitches aren't going to get us there, nor are the extractive dreams of Silicon Valley.

Industry 4.0 has not gelled into a genuine new paradigm of production, with new ideas and increases in productivity. It has yet to approach anything like the "American System," "mass production," or the "Toyota production system" did in their days—transformative aggregates of people, machines, and techniques that changed how people work and transformed the material world.

Invention and new technology are insufficient on their own. Universities and national labs brim with inventions and unlicensed patents; innovative startups go out of business every day for lack of adoption and lack of capital.

What's needed is a cultural movement aimed at transforming systems, from manufacturing to energy, mobility, and food supplies, a new Industrial Enlightenment. To do that requires some sense of what's next, and young people inspired to lead us there.

"My generation was trained to climb ladders," a young, highly educated energy entrepreneur told me, "not to build things." To retrain that generation, and the one after it, to actually build and maintain the industrial systems to support us, requires connecting that work to tomorrow's most pressing problems.

What are we building toward? What are the promising ideas to propel industry forward? What is possible that is not yet visible? The conversation is early, but this book aims to reframe how to talk about industrial futures.

ON COUNTING REVOLUTIONS

It has become popular to count industrial revolutions. Books published today on the future of industry feel compelled to recount the first, second, third, and fourth industrial revolutions—some going into five and six even. Klaus Schwab epitomized the trend in his "Fourth Industrial Revolution" and its buzzword "Industry 4.0."

As an engineer and a professional historian of technology, I don't count revolutions like software updates. I'm not even that comfortable with the single Industrial Revolution, with its implied analogies to sudden violent political overthrows, a term coined decades later to describe a process that itself unfolded over the course of a century. But Industrial Revolution has acquired a popular meaning that eases entry into important subjects, so I use it, sparingly.

The popularity for counting revolutions, however, suggests how deeply historical is our view of industry, and how equally deep run tensions about how to deploy that history when imagining the future. In what follows, we're not counting revolutions but rather looking at when the world went from no industry to industrial society (from zero to one if you are counting).

This is not a book about policy—though policy we need. It does not catalog statistics, nor does it end with a call for more R&D funding, nor more federally funded manufacturing pilots, nor improved apprenticeship programs, nor a revamping of engineering education, nor tax breaks for low-carbon technologies. All of those things are necessary, and numerous experiments are underway.

Rather, this is a book about the human dimensions of industrial improvement. It is about people who pursue that improvement like musicians play music, because it is a form of self-expression and they cannot do otherwise.

This book is about the interconnectedness of work, industry, culture, individual action, and collaboration. It is about the relationship between thinking and doing, and the relationship between the world of ideas and the world of things. It is about class conflicts, hot metal, linseed oil, pinched fingers, AI, and dung-sealed pistons. It is about canals and mines and delicate manufactured tea sets adorned with ancient dancing nymphs. It is about industrial systems as extensions, even mirrors, of human society, and about how both can change together.

It is my goal that this book, while inevitably rooted in moments of the 2020s, speaks to the next generation of industrial development.

I approach this topic having worked two careers, in parallel, for thirty-five years: as an electrical engineer and as a historian of technology. I began at the Oceanographic Institution in Woods Hole, Massachusetts, building and operating remote and autonomous robots in the deepest parts of the ocean, exploring shipwrecks, archeological sites, and hydrothermal vents – from the ancient Black Sea to the battle of Midway.

As we learned how to operate robotically, new ways of work were emerging, in a host of similar arenas. When I moved to MIT, my research focused on the human relationships with machinery, in undersea exploration, warfare, computing, aviation, and spaceflight. I've always worked in industry, too: as an inventor and consultant with a few dozen patents in autonomous helicopters, unmanned vehicles, and radio frequency navigation. For the past decade I've been a company founder and CEO, investor, board member, and mentor.

My work has always taken on some of the great stories of the technological world—ironclad warships, computing and automation in World War II, the Apollo lunar landings—and updated and reshaped them for today. The unifying principle has been human relationships with technology, in all of their dimensions and implications. My day jobs engage with engineering practice and system operations; in my spare time I fly airplanes, build electronics, write embedded code, and play electric guitar.

Though I consider myself a humanist, I have grown tired and wary of the humanities that offer plentiful critical theory but little synthesis, creativity, or optimism. Similarly in technology I am impatient with engineers who attempt to solve social problems with purely technical solutions, or with utopian technocrats who think only more technology will save us. This book seeks to imagine industry in ways that avoid these two extremes.

My previous books have been about the nineteenth, twentieth, and twenty-first centuries, leaving me a relative newcomer to the eighteenth. Yet as I found myself drawn to that very period when "industry" acquired its modern meaning, connections emerged that are beyond the concerns of more native scholars.

One reason a book like this can be written is the abundance of research and primary sources on the Lunar Society and its members that have emerged over two centuries. While the text draws lessons toward the end, the most valuable insights will be those the reader takes from the texture of the narrative.

A STEAMPUNK FAIRY TALE THAT
ACTUALLY HAPPENED

Let's look at the formative moments of industrial life: in the age of the Lunar Society, a group of inventors, entrepreneurs, scientists, and thinkers drove the Industrial Revolution. The group began with Matthew Boulton and his friend the physician Erasmus Darwin (grandfather of the great naturalist Charles Darwin).

In a series of monthly meetings and extensive correspondence beginning in the 1750s and continuing into the early 1800s, these thinkers exchanged and developed the big ideas that propelled the Enlightenment into the Industrial Revolution. They called themselves the Lunar Society because they met on nights with full moons so they could travel home safely (and they sometimes referred to themselves as "lunaticks"). They consciously brought the ideals of the Enlightenment to the industrial arts—what historians have dubbed the "Industrial Enlightenment," a special combination of creative thinking and making, "useful knowledge."

The Enlightenment was a far-reaching cultural, intellectual, and even religious movement in Western Europe in the seventeenth and eighteenth centuries that laid the conceptual and cultural underpinnings of our modern world. Though not necessarily centered in England, following the Scientific Revolution led by Isaac Newton, the Enlightenment posited that improvement in the world, and in human happiness, would progress through human activity and reason, more than through divine intervention, superstition, or royal patronage. The goal was apparently, though only partially, achieved through the American and French Revolutions.

The Industrial Enlightenment brought these ideas to the material world, to making and moving things. Through industry, enlightenment transformed into a credo of progress that advances in technology ultimately evolve toward perfection of human society and happiness.

The Enlightenment has always had its critics among historians and philosophers—from Karl Marx to Michel Foucault—focusing on the extremes of reason and capitalism, its social exclusions, and those left behind. A century of labor strife, the tragedy of World War I, the industrial slaughter of the Holocaust, atomic weapons, and environmental devastation of the twentieth century showed that progress in engineering and science do not necessarily move toward moral or human progress. Scholars have recently argued that "we have never been modern," in that politics, mysticism, and mythology have always underlaid enlightened institutions. Perhaps the Enlightenment could use a little defending and updating.

Historian Roy Porter argues the Enlightenment is best understood not as any one thing but as "a cluster of overlapping and interacting elites who shared a mission to modernize," which suits our stories here. Those elites shaped industrial society, with its values of "tolerance, knowledge, education and opportunity."

In today's pivotal moment, the stories of Watt, Boulton, and their Lunar Society colleagues offer new, yet sober, takes on common themes. Looking back to when it all began in eighteenth-century Britain helps discern contours of the coming transformation.

This book retells stories from this earlier era for a modern purpose: to illustrate how ideals and choices from the Industrial Enlightenment shaped the first Industrial Revolution, trace how those ideals led to today's systems, and encourage similar (though updated) ideals and thinkers to drive a new transformation. The aim is to revisit and rethink industry to help ignite a new Industrial Enlightenment to point us forward.

What historian Jenny Uglow said about the Lunar Society also applies to the generation of young industrial leaders rising today, who I hope are reading this book: "together they nudge their whole society and culture over the threshold of the modern, tilting it irrevocably away from old patterns of life towards the world we know today." Retelling these stories with an updated lens refashions the Lunar Society into the original steampunk fairy tale, one that actually happened.

SHADOWS OF INDUSTRIAL ENLIGHTENMENT

Historians try to see the past on its own terms, not solely through the lens of the present. Yet patterns today resonate with the past, and those harmonies clarify what is fundamental and what is new. Shop floor struggles, management of knowledge, supply chain challenges, and international competition are fundamental. High-capacity batteries, real-time sensing of production processes, computer-controlled additive manufacturing, remote work, and generative AI are new.

A journalist recently interviewed me on the history of manufacturing. When I explained how it helps understand what's fundamental about industry, he misquoted me as calling for "a return to a more traditional way of doing things." That is not what I said, and that is not what this book is about. The future will be unlike the past. The past helps us understand that what was once possible may be possible again. Its very alien nature helps us see today's world anew.

The aim is not to heroize or idealize the Lunar Society nor its members—in fact, their limitations are as instructive as their insights. The Industrial Enlightenment missed its lofty goals in plenty of ways. Some say these are inherent faults of capitalism. The evidence here suggests they were more the results of avoiding difficult realities, optimizing for some goals over others, and limited human vision.

Industry's problems, too, trace back to the original Enlightenment thinkers and how their Industrial Enlightenment failed to address two problems: environment and work. Factories polluted towns as quickly as they arose. Early industry was not always able to offer workers dignified jobs and shared prosperity in exchange for their contributions (though sometimes it did).

The Industrial Enlightenment began a process of extracting resources from the earth in order to add energy to industrial systems, reaping enormous gains in productivity. Now the crisis of climate change has forced itself on industry and forced us to question that model. Yet to transform our systems, along with any shift away from fossil fuels, factories will need to make a whole lot of new stuff, from batteries to cars, pipes, and electrical connectors. Labor relations were already in flux before the pandemic, and now work—and the culture, politics, and technology that surround it—is transforming again.

So how to imagine futures for industry that get the two problems right: environment and work? What does industry look like when it strives to optimize for the lowest carbon footprint as well as the greatest profit? When it values resilience as much as efficiency? When it upholds dignified, inclusive, sustainable work?

It is easy to imagine inventions and breakthroughs that might save the day. An entire media industry amplifies the self-interested claims of inventors and investors that their widget or code will inevitably change the world—what some have called "innovation speak."

But utopian visions ignore the hard work of transformation: refinement, adoption, learning, maintenance. Not to mention darkness and failure. These and other slow, human processes characterize industrial change, and leaders and engineers must take them seriously to avoid chasing after the next shiny thing. How can we talk about futures for industry that account for these human realities, hopeful but not utopian? How can we enroll a generation of people in the human and material satisfactions of this work?

The pages that follow sketch an outline. Completing the picture should be left for the new industrialists: builders, thinkers, and entrepreneurs, already at work reinventing industry. The Lunar Society shows how these young people inherit traditions of thinking doers whose industrial transformations drove social change. These stories suggest guidance for imagining industrial futures in ways that are both realistic and human-centered.

BIRMINGHAM AWAKE WITH TOYS

Our story begins in the middle of the eighteenth century, in a Britain rising to the height of its imperial powers. Britain defeated France in the Seven Years' War, which in hindsight one could accurately call World War Zero. The 1763 Treaty of Paris gave Britain dominion over Canada and much of North America east of the Mississippi, including Spanish holdings in Florida and the Caribbean.

The British Navy and its wooden walls protected the island nation and extended a newly coherent empire, supported by a thriving Atlantic trade in raw materials, finished goods, and enslaved human beings. The royal dockyards had become the largest manufacturing sites in the nation, using people and craft skills to fuse oak, canvas, rope, iron, and bronze into complex machines to project national power. Enslaved workers on plantations harvested sugar in the Caribbean and throughout the colonies, sweetening the diets of the English middle class. And commerce the empire needed, for the wars had imposed a large debt that London would try to service by taxing the colonies.

King George III ascended to the throne in 1760, at age twenty-two, himself a child of the Enlightenment, a student of science and astronomy. He amassed a book collection so foundational it formed the basis of the British Library, where you can visit it today. The London that Americans picture as their enemy in the Revolution (as well as the one they visit today as tourists)—wealthy, baroque, coiffed, and a little decadent—dates largely from this period, including Buckingham Palace, bought by George III in 1762 and renovated for his new wife, Queen Charlotte.

Birmingham, a hundred miles to the northwest of London, had little of that Georgian flourish. It didn't even have a proper road. Barely a town when the century began, the city was literally off the beaten path, with no direct route to London.

Yet the place buzzed and grew quickly. Birmingham tripled in size in the first half of the century, and then tripled again before 1790, to become the third most populous city in England. It was a city on the move, and a city of makers. More recently, Birmingham has given the world cultural icons like J.R.R. Tolkien and Led Zeppelin's Robert Plant. But in the eighteenth century it gave the world industry and toys.

"I had been among dreamers, but now I saw men awake," William Hutton, a bookseller, wrote of his first visit to Birmingham in the early 1740s. "Every man seemed to know and prosecute his own affairs." He found the city's people "full of industry." Before factories and steam engines, industry was a virtue that drove people's behavior.

Birmingham had a metalworking and smithing tradition, beginning with nails and ordinary tools. The city's artisans gradually worked their way up the value pyramid to make fancier stuff: buckles, buttons, guns (often destined for the slave trade), jewelry, watch chains, and snuff boxes. These small metal products came under the term "toys."

When London royalty began wearing diamonds on their clothes, Birmingham metalworkers figured out how to cut steel studs and polish them to a nearly comparable shine—diamonds for the aspiring and middle classes. Such toys were an ideal industry for a provincial outpost. The small, light objects required modest amounts of capital, raw materials, and transport networks.

By 1750 about twenty thousand people worked in the toy trade, much of their output going to export. Birmingham would remain the metal manufacturing center of the world for more than a hundred years. Though the rapid growth and pollution would give the city a wearied, gray countenance, visitors came from around the world to wonder at its factories. The streets thumped and clanged with the music of industrial work.

Despite the humble products, the toy trade drove Birmingham manufacturers to flexibility, productivity, and low cost. Fashionable pieces were popular with consumers, but fashion changed quickly, and makers needed to adapt with speed. "This offspring of fancy, like the clouds, is ever changing," Hutton remarked. "The fashion of today is thrown into the casting pot of tomorrow." As today, seemingly insatiable consumer demands for fashion and status drove industrial transformation.

DESIGNED IN CALIFORNIA . . . MADE WHERE?

Today, when you open up any Apple product, you will see a label that reads "Designed by Apple in California." This statement contains an implicit value judgment, intended to associate three words (design, Apple, California), drawing in the aesthetics of the product with the cultural iconography of the place: Steve Jobs as the hipster James Watt of the Information Age (though, as we shall see, Jobs resembled Josiah Wedgwood more than James Watt).

The phrase, "Designed by Apple in California," is notable for what it doesn't say: where the device was manufactured. That's printed elsewhere on the product, by law, in much smaller text. It also doesn't say where the pieces came from. Likely workers in China or Vietnam assembled the product with parts from around the globe. The value judgment implicit in "Designed by Apple in California" is that what matters is where the design occurred—the creative, innovative part—and that the manufacturing location is irrelevant to the consumer's enjoyment of, or identification with, the product.

"Design here, make there" was a 1990s mantra, emerging out of the neoliberal free trade era that shaped the past decades. Buried in this simple phrase is Apple's assumption about technologies around the millennium: product innovation is where the value is added, in the new phone design, the new operating system, the new app. Manufacturing—how one sets about actually making these things in the millions or hundreds of millions—is secondary, for someone of less status in another country to worry about. For at least the past generation, process innovation in America has been distinctly a rung down on the creative ladder below product innovation, as hardware has been lower status than software.

American companies have come to separate product innovation from process innovation. This divorce stemmed from visions of a post-industrial future where knowledge work overtook making things as a source of value and prestige. It had historical roots in America's manufacturing strength in the mid-twentieth century.

After World War II, the United States, secure in the astonishing capacity of its factories that won the war, invested in research and development. Here it was supported by "the ideological and methodological limitations of mainstream economics," which failed to see the close links between producing knowledge and producing things.

The document that supposedly laid the blueprint for postwar US innovation policy, Vannevar Bush's famous report, *Science the Endless Frontier*, mentions manufacturing only once. The later era of neoliberal economics furthered the divorce: economists, venture capitalists, and Wall Street analysts elevated the "core competency" of a company. Everything else could be modularized and contracted out, preferably to another country where one didn't have to think about or see labor conditions.

Startups, in particular, have been able to raise capital for product development but then encounter barriers when they need to scale up manufacturing: finding talent, suppliers, and financing. Often, to further scale, they move their operations overseas. In sector after sector, including most recently solar panels, offshore wind, batteries, and electric vehicles, the US has developed advanced technologies and then ceded manufacturing leadership to other countries.

The trouble is, as our journey into the Lunar Society will show, product innovation and process innovation are not so easily separable. In fact, they're inextricably linked. It's difficult to retain an R&D lead when one loses the links to production. This holds particularly strong for complex, high-value products at the cutting edge of technology and design. In the early twenty-first century, Chinese companies began manufacturing such products designed elsewhere, and quickly mastered their processes. Before long, smart engineers in China, like the Birmingham metalworkers, began seeing how to improve the processes, what was possible, and then how to improve the products themselves.

Consider Moore's law, the best-known description of technological progress in the past half century. What is Moore's law but a statement about process innovation? That progress in computing (and mobile phones, networking, AI, and so much else) is driven by progress in manufacturing? Indeed, when the Taiwan Semiconductor Manufacturing Corporation (TSMC) began fabricating chips for so many companies through its foundry model, it quickly came to own the frontier in semiconductor processes and to share those improvements first with its largest product design customers like Apple. Less evident is that Apple's model was, in practice, not a "throw it over the wall" (or across the ocean) type of model, but one that required hands-on interaction with numerous Apple engineers living for months at a time near the factories in Shenzhen where iPhones are made.

FABRICATING THE PIN FACTORY

The greatest improvement in the productive powers of labour,
and the greater part of the skill, dexterity and judgment with
which it is anywhere directed, or applied, seem to have been
the effects of the division of labour.
—Adam Smith, *The Wealth of Nations*, 1776

The problem began at the birth of economics, when Adam
Smith famously described the manufacture of pins. One person, Smith wrote, working on their own, could make perhaps
twenty pins in day. But in Smith's account, the division of
labor separates the process into eighteen steps: drawing out
the wire, straightening, cutting, sharpening the point, putting
on the head, and so on. People specialize, getting extremely
good at only one of these smaller operations. "It is even a trade
by itself," wrote Smith, "to put them [the pins] into paper."
In this scheme, ten workers (all men in Smith's description)
could make tens of thousands of pins in a day—a huge gain
in productivity.

The few pages describing pin manufacture have become
the most famous passage in economics. That they appeared in
easily digestible form at the very beginning of the modern discipline's founding text, *The Wealth of Nations*, surely improved
the anecdote's popularity (followed by many hundreds of less

evocative pages on tariffs, international trade, and comparative advantage). Ironically, the pin factory story is the only place that Smith discussed the industrial changes taking place in his time.

Yet Smith never visited a pin factory. He also didn't visit the more creative factories run by his friends nearby. He drew the account from French sources, including Denis Diderot's *Encyclopédie, ou Dictionnaire Raisonné des Sciences, des Arts et Des Métiers*, and embellished and misinterpreted them.

Pin factories of the day were well documented in France: workers actually divided into fewer than ten trades, not eighteen. Specialized skilled workers did not dominate but worked side-by-side with unskilled ones. Nearly half of the workers were women. Smith did not realize that the economics of the division of labor in pin making depended mostly on wage differences between sexes.

Smith suggests that the division of labor could continue ad infinitum, continuously improving productivity. But further divisions of labor had been tried and did not improve output. More specialized labor was not always more productive.

Pin factories that did not adopt the division of labor were potentially one hundred times more productive than Smith reported (making the new organization he described only two times more productive, not 240 times more). Larger factories were not more productive than smaller ones. It was left for industrial philosopher (and computer inventor) Charles Babbage to describe, decades later, that the productivity improvements in pin making depended on different wages for the different workers.

The division of labor became the software, even the AI, of its day—an animating principle that drove industrialization. Smith's errors and fabrications need not indict his core ideas nor the benefits from the division of labor (a term he coined). But they illustrate a larger point that has plagued economics in its relationship to manufacturing ever since: an emphasis on the neat theory, the digestible point, the generalizable principle, over grounded social and technical relationships of how things are actually made.

The word "engine" appears in *The Wealth of Nations* to describe mercantilism and monopoly as engines of the economy. The text never uses "engine" to describe the machines that Smith's own friend, James Watt, had devoted his life to perfecting for ten years before Smith wrote. Smith never visited (nor mentioned) the Birmingham toy factories nor Matthew Boulton's innovative Soho works, though they were all in operation for decades before *The Wealth of Nations* appeared.

Smith's distortion matches surprisingly well how today's economists missed the social, political, and technological importance of manufacturing in favor of all-out globalization. "The central cause of the profession's failure," writes economist Paul Krugman of the 1990s and early 2000s, "was the desire for an all-encompassing, elegant approach that also gave economists a chance to show off their mathematical prowess." In a more general sense, writes policy expert William Bonvillian, "whether mainstream economics can account for the critical role manufacturing plays in the U.S. economy has come into question with increasing frequency." Nobel-winning economist Angus Deaton criticized modern economics for its "limited range and subject matter," "unmoored from its proper basis," in human welfare. Focused on supply and demand, efficiency, and market equilibria, economics has sought to understand the effects of technology on economic growth—but all while still seeing technology as an exogenous variable, an impersonal historical force that comes from outside society.

"The details of America's innovation system and its weaknesses are essentially alien to macroeconomics."

MIT political scientist Suzanne Berger, who studies manufacturing from the ground up, has argued against economists' industry agnosticism, in which they mix up manufacturing with agriculture. While agriculture experienced productivity growth and an ever-shrinking labor force, productivity in manufacturing has been slower to grow, with consequently less decline in the labor force. Thanks to the misreadings of Smith's descendants, economic policy in the United States enabled the decline of manufacturing, confident that the new information and service sectors would replace those jobs.

This blindness had crippling policy implications. America created an economy in which high growth is concentrated in software, healthcare, advertising and marketing, and similar services. An "innovation first" model has dictated that this emphasis—the less material the better—is the key to a modern "service economy."

The trouble is, legacy sectors still make up most of the economy: utilities, civil construction, building, agriculture, mobility, education, mining, and the like. These sectors tend to have established structures, vested interests, high regulation, and high barriers to entry.

High on this legacy list is manufacturing. Manufacturing is important, not necessarily because nations should build everything themselves nor because it makes up a huge fraction of the labor force (less than 15 percent in 2020, at under thirteen million workers, down from a high of almost twenty million in 1980).

But manufacturing is a unique component of an industrial economy. It has a high multiplier effect. Every dollar of manufactured goods adds almost four dollars to the larger economy, and every manufacturing job accounts for about three and a half jobs in related industries. Every manufacturing dollar and job engenders additional dollars and jobs in logistics, delivery, food service, energy, distribution, sales, support, and other sectors. Manufacturing thus indirectly accounts for about one-third of gross domestic product and employment. Manufacturing jobs, like factories themselves, are relatively steady, often free of the schedule chaos that characterizes other sectors like retail.

Few argue that the United States should compete in making low-cost commodities. For advanced products and processes, however, the situation is different. Manufacturing connects process innovation to product innovation: the value of designers working near manufacturers to quickly resolve production problems. Manufacturing is itself a major source of improvement and new ideas. Manufacturing commodity items like small mobile phone batteries, for example, informs how we can make batteries for electric cars, and even how to make the cars themselves (which are more or less batteries that drive around).

By contrast, China has succeeded not solely because of lower labor costs but because it has built an ecosystem that links process innovation and product innovation: "simultaneous management of tempo, production volume, and cost, which enables production to scale up quickly with major reductions in cost. . . . even in industries that are highly automated."

"Produce there/innovate there," Bonvillian writes about China's rise, may have been even more disruptive than "innovate here/produce there."

A few years ago I spoke with a senior executive at a major European aircraft manufacturer, a manager in charge of hundreds of factories around the world. A modern jetliner is a capital-intensive, highly standardized, tightly regulated product. It is a product that has stayed basically the same for the past thirty years, and an enormous backlog of orders means they'll be making nearly the same jetliners for at least the next twenty.

Yet when asked to identify his greatest challenge, the executive said, "Organizational learning." The global manufacturing system was so dynamic—with constant changes in supply, qualified vendors, tariffs, wars, politics, and many other variations—that even while making a standardized, regulated product, this large organization needed to be learning all the time. A seemingly innocent change in which vendor makes something as simple as a rubber gasket can result in defects if the material behaves only slightly differently. Boeing's persistent problems with production painfully highlight how difficult it can be to make the same thing over and over again, when that thing is an expensive, life-critical assemblage of high-performance parts.

Manufacturing is in constant conversation with the dynamism of the world. With changing supply chains, changing workers, changing market demand, new technology, the factory that can be set once and make the same thing the same way forever does not exist.

Far from a dirty, antiquated legacy sector, manufacturing is the lifeblood of industrial societies. It is a noble enterprise, humbly engaged with the complex material realities of the world. It is also a hopeful enterprise, creating the things that human beings need for survival. While deeply material, it is also driven by ideas and inspires new ideas of its own. In the United States, it represents the middle economy, a path for people with skills today to improve and grow into career pathways.

Moreover, manufacturing will be critical to address climate change.

The United States is seeking electrification as a source of decarbonization in numerous sectors, yet the transformers that link electric grids are largely made overseas. Even those that are made domestically depend on specialized steels imported from China. Electric utilities are simultaneously faced with climate-induced disruptions (for which they must stockpile transformers) and increased demand due to electrification.

Manufacturing is also crucial for national defense, and those manufactures—in forging, shipbuilding, and advanced semiconductors, for example—have been hollowed out as well.

From the time of the Lunar Society, if not before, manufacturing has always been tied up with questions of international trade, supply chains and transportation, labor, skill, and expertise. Breakthroughs are overrated. Most technology evolves through incremental, consistent advances of the kind closely linked to manufacturing.

Modern manufacturing originated in the idea of Industrial Enlightenment, that our greatest ideas are expressed in *how we make things*. Thus, *how we make things*—and market, distribute, and use them—can be a vehicle for the improvement of human life, both by supplying our basic needs and by catalyzing social change.

"MASTER OF EVERY METALLIC ART"

Few can claim, as Matthew Boulton could, to have been present at the founding moments of the Industrial Revolution, to have instigated key elements, and to have imagined it in advance. Engaging, social, and matter of fact, Boulton possessed high technical skill in his trades and the humility to know when excellence had come to him. His curiosity both begat the Lunar Society and enabled him to see the future of industry. Entrepreneurial drive coupled with a lack of pretension gave him an eye for fashion and an ability to sell his products to the social aspirations of consumers. He dressed kings and queens and helped common people dress like royalty.

Boulton became wealthy through business and marriage, but he never tired of industry, science, and experimentation. A better entrepreneur than a capitalist, Boulton ultimately compromised the scale of his fortune through expansive (and expensive) intellectual and business adventures. Perhaps more important to him, he led an interesting life, a nexus of enlightenment, charisma, industry, and power.

Unbeknownst to Adam Smith, before *The Wealth of Nations* Boulton's manufactories refined the division of labor with exacting management and precision, as well as power and machinery. He improved metallurgical and process technologies and pioneered industrial methods of manufacturing silver plate, jewelry, buttons, and sword hilts. His silver work, coins, and metal products are among the finest examples of their times—they remain on display in museums (including the Victoria and Albert in London). In ormolu (a term Boulton may have coined), a kind of gilded gold-plating, he became the largest and leading producer, and he led as well in polished cut-steel buttons and jewelry. Later in his career he created the steam-powered minting of coins.

Boulton was born in 1728 (most members of the transformative Lunar generation arrived in the few years around this date), the son of a typical Birmingham toy maker. When Boulton took over his father's business at a young age, it prospered. He enjoyed company and gave generously of his enthusiasms. Lunar Society member James Keir later wrote of Boulton: "Mr. B. is proof of how much scientific knowledge may be acquired without much regular study, by means of a quick and just apprehension, much practical application, and nice mechanical feelings." Boulton, Keir said, "had very correct notions of the several branches of natural philosophy, was a master of every metallic art, & possessed all the chemistry that had any relations to the objects of his various manufactures." Boulton was also an optimist—keeping his cheerful disposition in the hardest of times, a trait that would crucially offset the gloomy James Watt.

Boulton continually sought to broaden his domains, and this endeavor was social as much as experimental. He married Mary Robinson, daughter of a prosperous textile merchant. Her dowry brought him capital for his business. The marriage also connected him to the Robinson family's physician, Erasmus Darwin.

Darwin, a portly, boisterous poet, naturalist, and physician, lived in Lichfield, just a few miles up the road from Birmingham. Educated in Cambridge and Edinburgh, he enjoyed higher status than Boulton. Darwin penned botanical treatises framed as epic poems, and he wrote extensively on his idea of a "fiery chariot"—an early idea for a steam locomotive.

Both promising young men in their twenties, Darwin and Boulton became friends in the 1750s. Both had married well and were on the move, making their ways in the world. Darwin quickly established himself as socially and professionally prominent. John Whitehurst, an instrument and clockmaker who possessed the height of mechanical craft and expertise in core technologies of the industrial factory, joined the trio as well.

In 1758, Darwin published two letters in the *Philosophical Transactions of the Royal Society* on heat in vapors and their relationship to electricity. While of minor scientific importance, these experiments—which Boulton may have assisted—kindled Boulton's interest in electricity.

"We know it's matter," he wrote in 1755 of the mysterious fluid, "& there is wrong to call it Spirit." This snippet from Boulton's early notes on electricity hints at his deep empirical approach, as well as a desire for material, as opposed to divine, explanations.

Part of what makes the Lunar Society members so interesting and attractive to our inquiry is that their industrial mindset had not yet narrowed into a humorless utilitarianism. Later industrialists (including Boulton and Watt's own children and the later Charles Babbage) would evolve toward a calculating managerial mindset, interested primarily in seeing the world through measurement and numbers. In response, the Romantic era's reaction to industry would elevate feelings and nature.

For Boulton and Darwin, mind, hand, and heart had not yet split. Theirs was a sentimental empiricism—relying on human experience and senses as key instruments for scientific observation. They did not limit themselves to industry but experimented in life as well—how to bring modern principles to education, family arrangements, and the home.

Boulton collected books and conducted experiments on materials and their properties (electrical and thermal). He felt himself better able to approach electricity than older philosophers, "for we can both hear it and see it smell it & feel it . . . let us consider it just as it appears to our senses." This was an era with no volt meters (nor even any volts): experimenters like Boulton measured electricity by the amount of pain that shocks inflicted on their bodies.

This interest in electricity led Boulton to Benjamin Franklin.

Both Boulton and Franklin bridged worlds of making and thinking. Though less fortunate in his circumstances than Boulton, Franklin had made his way as an artisan—a printer, a cutting-edge technical and cultural trade of the day. Twenty years older than Boulton, Franklin sold his business in 1748 and achieved great wealth at age forty-two. Ever seeking social improvement, he energetically pursued a path toward gentlemanly repose. For him, like Boulton and other members of the Lunar Society, gentlemanly status meant scientific experiments with electricity.

Though Franklin's electrical work covered only a few years, he quickly consolidated the field in experiment and theory. In 1752 he showed that electricity and lightning were equivalent. Though the iconic kite and key experiment is an oversimplification, Franklin did show that the modest laboratory electric effects that had engaged scientists so far (despite their pain) were actually aligned with the most powerful forces of nature.

Franklin coined the terms *positive* and *negative*, *battery*, *charged*, *neutral*, and *conductor*—still core vocabulary in our language of electricity. His law of conservation of charge remains fundamental to our understanding. "He found electricity a curiosity," noted one Franklin biographer, "and left it a science."

Electricity runs in the background of the Lunar story, indeed of Industrial Enlightenment: a curious phenomenon of nature, confounding current knowledge but amenable to experimentation. Franklin's lightning rod is described as the first practical invention to come from Enlightenment inquiry, but it would be a generation after the Lunar Society that electricity itself became an industrial tool. After yet another century, the tiniest pulses of electricity in the nearly pure crystals would further transform the world: "semiconductor" is still half Franklin's word.

By the time Franklin arrived in England in 1757 to represent the Pennsylvania assembly, he had been elected to the Royal Society, the "invisible college" founded by Robert Boyle and others that congealed the Scientific Revolution in the seventeenth century. Franklin was the most famous American scientist in Europe, and possibly the most famous American. Around this time the Scientific Revolution was expanding from a metropolitan movement based at the London Royal Society to a more provincial one based in philosophical societies.

Boulton and Darwin, following Franklin's latest developments from their provincial outpost, became fans. In 1758 Franklin came to Birmingham to visit family and his friend the printer John Baskerville (creator of the Baskerville typeface and later a Lunar Society member). On his tour Franklin visited a silk mill, new canals, and workshops. He called on the dissenting clergyman Joseph Priestley, who was then writing a book, with Franklin's help, on the history and state of electrical research, which would prominently feature Franklin's work and bring it to broad European audiences.

Franklin also met with Darwin and Boulton and shared some friendly lab work. Boulton had already been interested in electricity, but his work with the famous American now made him the local expert. Franklin returned to Birmingham in 1760, when he and Boulton continued their sociable experiments.

Boulton had a tough year in 1759: his father and his wife both died. Despite the grief, he quickly married his wife's sister, Anne Robinson, doubling his capital through her inheritance and enhancing his access to others'. Her brother passed shortly thereafter, further amassing Boulton's trove. Though inherited wealth and family connections helped Boulton get started, his was a hybrid fortune, on the border between gentry and commercial. Like Franklin, he was simultaneously pursuing commercial success and social advancement.

Despite his windfall, Boulton remained hungry. Soon after his second marriage, he began purchasing the land to build his new manufactory, later called Soho. He tore down an existing rolling mill and replaced it with a new one for finishing metals. Part of the motivation was access to power: the site was slightly outside of town but near a source of water. Later, the power from the millpond would prove both inadequate and seasonal; it froze in winter and dried up in summer.

Soon Boulton was employing hundreds of workers at Soho, and nearly a thousand a few years later—the largest manufactory in the Western world. The company always struggled financially, in part because of Boulton's ever ambitious plans, but it became a national center for manufacturing.

Unbeknownst to Adam Smith, Boulton employed the latest techniques of management and machinery, exploiting the division of labor to create specialized tasks and deploying water-powered machines wherever possible for polishing and lapping metals to make finished goods of high quality. When he visited the factory, Darwin observed "a large waterwheel in one of the courts," which drove "motion to a prodigious number of different tools." A later catalog lists more than a thousand designs for buttons, buckles, hilts, chains, and other luxuries.

Boulton obsessively courted prestigious visitors as potential customers. A commission for vases from King George III and Queen Charlotte in 1770 capped Boulton's pursuit of prestige customers, with positive spillover for the consumer business.

Industrial tourism was already beginning in Birmingham, but Boulton's SoHo works soon eclipsed the button, watch, and gun factories as the main attraction for the well-heeled. Soho had a grand entrance, a showroom, and numerous small rooms, "which like Bee hives crowded with the Sons of Industry."

Visitors included ambassadors from the courts of Europe, the naturalist Sir Joseph Banks, astronomer Sir William Herschel, actresses and artists, Lord and Lady Hamilton (and her lover Lord Nelson), dukes and duchesses from the continent.

Boulton and his wife entertained guests over dinner at their home next to the factory. "Mr Boulton himself is the most interesting," wrote one aristocratic visitor. "We passed an evening listening to him, & we wished he had never ceased to speak."

Also infiltrating were no small numbers of spies and industrial plagiarists—conflicting Boulton's Enlightenment ideal of sharing knowledge with protecting commercial trade secrets.

Despite its prominence, Boulton's toy business perennially lost money, due to servicing the debt on the massive new factory, uncontrolled product variation, and lapses in management, pricing, and receivables. Boulton had a tendency to start grand new ventures rather than refine those at hand.

All this was just beginning in 1765, when Boulton, Darwin, Whitehurst, and Franklin began talking and corresponding about electricity, meteorology, and the fire engines then in use to pump water out of mines. Historian Robert Schofield calls this early group "The Lunar Circle" as a prelude to the more organized Lunar Society. Each friend straddled the worlds of mind and hand; each absorbed, consciously or unconsciously, Enlightenment empiricism and practical knowledge. These men shared a sense that exchanging knowledge and ideas were social activities, best done through correspondence, friendship, and association.

And knowledge they needed, for the world was changing quickly in Boulton's competitive businesses.

The Lunar Circle members were provincial, free of big-city London judgment, but also hungry and striving. Though he would forever stand in a middling social class, Franklin himself abandoned science and invention, realizing that in his present world diplomacy and leadership (particularly in Europe) were paths to greater prestige. "In the eighteenth century, artisans and mechanics," writes Gordon Wood, "all those who worked for living, especially with their hands, no matter how wealthy, no matter how many employees they managed, could never legitimately claim the status of gentlemen."

Schofield notes an article from 1803 that mentioned "English Worthies" under headings as "Ministers and Statesman, Lawyers, Judges, Divines, Voyagers, Travelers, Mathematicians, Naturalists." The list makes no mention of either inventors or manufacturers. Even by then, nearly half a century later when the Industrial Revolution was well under way, English heroism derived from military, political, aristocratic, even professional status. It remained a radical idea that mechanical and commercial nerds could shape the world as much as warriors and kings.

INDUSTRY

Your image of manufacturing might be Henry Ford's Model T line in Highland Park, or a modern automotive assembly line cranking out thousands of vehicles per day, or even a squeaky-clean chip factory.

These are indeed important industrial sites, but they are better conceived as nodes in a system of material flows rather than simply the place where things are *created*. An assembly line (more precisely called a *final assembly line*) is more than a place where cars originate; it is a place where thousands of parts come together to form a car, each arriving after traveling through their own fabrication and logistics networks.

Each assembly plant at any car maker you have heard of draws on dozens of sites from Tier 1 suppliers (who might make the body panels or brake lights), on hundreds of Tier 2 suppliers (who make smaller assemblies), and even more on lower levels.

When completed cars emerge from the factory, they reenter transportation networks for distribution, finding their way to your local car dealership. The same is true for computer chips or clothing. For this reason, manufacturing is inseparable from logistics, mobility, energy, and other industrial systems. And those systems rely on trucks, tires, wires, connectors, generators, and other manufactured things to keep running.

Your entire relationship with a product—whether a few minutes in the case of a piece of fruit or years in the case of a car—is also only one stop on its journey, after which it will enter still further systems of refuse, waste, and possibly reuse or recycling.

These networks become even more complicated in the case of electric vehicles (EVs). Far from simply putting a new type of motor into an existing vehicle template, producing EVs requires a transformation of supply chains, indeed a very reconception of what an automobile is.

A traditional internal combustion vehicle is an aggregate of steel, rubber, glass (from the time of Henry Ford), and more recently plastic and electronics. By contrast, "EVs represent a new class of cyberphysical systems," reads one recent expert analysis, "that unify the physical with information technology, allowing them to sense, process, act, and communicate in real time within a large transportation ecosystem."

EVs are roving software bundles being constantly upgraded. They are extensions to the electric grid, greedily consuming at certain times, quietly storing and generously offering energy at other times. They require new raw materials, new factory processes, new skills to build and maintain. The fuel tank (the battery) can cost as much as the rest of the car. EVs will induce new automation technologies and induce new consuming and driving behaviors. And culturally they carry the hopes of governments, activists, and consumers for a global energy transition—hopes sometimes in conflict with gritty engineering realities.

So far, I have used the term *manufacturing* to refer to the process of making things, and indeed it is a core subject of this book. But our story also draws in a broader term that better captures the breadth of what's at stake: *industry*.

The modern reader may associate industry with dirty cities, contaminated Superfund sites, or smoky factories. If you include the word in a prompt for a generative AI image, the resulting picture will include belching smokestacks. Indeed, industrial culture must be accountable for these degradations. Sometimes we colloquially use *industry* to refer to economic sectors: the financial industry, the entertainment industry, the software industries.

Here I'll use the word in a more traditional way to refer to industrial systems—that is, the large-scale technological networks that provide for basic material human needs: food, water, shelter, defense, energy, mobility, work, and social status. These systems exist in the physical world and are large, capital intensive, and slow to change—not least because when you supply something that is genuinely important like food or electricity, change is risky. Disruption is the enemy.

McKinsey consultants have labeled the US industrial eco-system "the Titanium Economy." They identify almost seven hundred publicly traded companies in this sector; 80 percent of them are in the $1 billion–$10 billion range, with two to twenty thousand employees. For privately held companies, industrials make up 38 percent of the top one thousand firms by revenue and are growing rapidly. The McKinsey folks believe the United States is capable of producing a thousand Teslas in every industry. But the popular culture of technology has yet to catch up. We hear endlessly about new laboratory innovations that might transform something someday.

Industrial systems often remain invisible, just below the substrate of modern life, until something fails. Disruptions like the COVID-19 pandemic have a way of making invisible systems shockingly visible again.

Actually, this use of *industrial*, as in *industrialism* and *Industrial Revolution*, is relatively new, dating to the nineteenth century. An older use is also relevant, indeed crucial, here.

Industry, in addition to referring to economic sectors, is also a human virtue: sustained application of effort. As William Hutton observed when he called workers in Birmingham "full of industry," one calls a person *industrious* to contrast their behavior with sloth and idleness. Although the word had been around earlier, this use exploded in the time of the Lunar Society, the eighteenth century.

In fact, *prior* to the Industrial Revolution occurred the *Industrious Revolution*—a change in how people, particularly in Britain, worked with new energy and purpose. Households both consumed and labored more in relationship with markets outside the home, enhancing both demand for goods and labor supply.

The British ritual called "tea," featuring the consumption of imported goods on purchased wares, spurring family labor to pay for them, emerged in the late seventeenth or early eighteenth century and epitomized these behaviors.

What set Boulton and Watt apart from the generations of metalworkers and makers who came before them? In addition to sheer effort and striving, the Industrious Revolution entailed an evolution from the natural cycles of the seasons and the fields to the syncopated rhythms of the clock and the machine. Boulton and Watt, while exceptional people, were also typical of their age.

A surprisingly large number of Lunar Society members were dissenters or nonconformists—technical language at the time for those outside the Anglican Church, including Quakers, Unitarians (Watt, Wedgwood, Priestley), Puritans (Franklin), Baptists, and many Scots. Well into the nineteenth century, dissenters were barred from public office, military commissions, and university education. Because of these restrictions, many turned to commerce and artisanship. Disproportionate numbers of industrial and commercial improvements came from members of these groups.

To educate these outsiders outside of Oxford and Cambridge, "dissenting academies" arose with more practical curricula in history, arithmetic, business, and experimental sciences than Oxford or Cambridge would allow. The leading such school, Warrington, about halfway between Manchester and Liverpool, linked a number of the players in our story. Joseph Priestley arrived there to teach in 1761 and modernized the curriculum.

Priestley believed that teaching history as part of a practical education "resembled the experiments made by the air pump, the condensing engine, or electrical machine, which exhibit the operations of nature." His *History and Present State of Electricity* (1767), intended as an introductory textbook, shaped the field for generations of researchers, including Alessandro Volta and Michael Faraday.

As if to symbolize the link between the flow of ideas and the flow of goods, Warrington opened the same year as the Bridgewater Canal, just over a mile away from the school. By comparison, Priestley believed Oxford and Cambridge to be "pools of stagnant water."

In a similar vein, Benjamin Franklin, a child of Puritans, nearly a generation older than Boulton, embodied an American version of Enlightenment ideals. He became known for his aspiring middle-class aphorisms like "Early to bed and early to rise makes a man healthy, wealthy and wise" and "Haste makes waste." Franklin's autobiography remains one of the most popular books in American history, and it stands as a kind of founding tale for American character and self-reliance. It was even cited by sociologist Max Weber to exemplify his theory of the Protestant ethic and the spirit of capitalism, "in almost classical purity" (though Franklin did not live as ascetically as Weber believed).

Franklin adapted Enlightenment values to American social progress. He used the phrase "industry and frugality" no fewer than thirty-six times in his short autobiography. The aspiring middle-class diligence of the printer, the metalworker, or, as we shall see, the instrument maker or potter laid the foundation for the Industrial Revolution as much as any technical invention did.

Boulton, Franklin, and their peers occupied a borderline position between artisanship and ownership, between the working and the leisure classes. This intermediate position also proves crucial to our current moment of industrial transformation. "By absorbing the gentility of the aristocracy and the work of the working class," noted the historian Gordon Wood of the American Revolution, "the middling sorts gained a powerful moral hegemony over the whole society." In contrast to white-collar economists and managers manipulating data and analytics, today's industrial entrepreneurs occupy a similar position of translation—between a blue-collar workforce and the managerial class—a crucial social and political nexus.

These middle-class values would be later be derided by all number of critics as humorless, mundane, and calculating. Franklin biographer Walter Isaacson called these critical attitudes "a snobbery that would come to be shared by very disparate groups: proletarians and aristocrats, radical workers and leisured landowners, Marxists and elitists, intellectuals and anti-intellectuals."

Despite the derision, when one talks to industrial employers today, they are less concerned about skills in robotics and AI and more concerned, as were Boulton and Watt, with finding workers who can show up and work sober for sustained periods. Franklin's habits of mind might perhaps be recovered and modified for this world. Trust, reliability, and promptness are valued in any collaboration, while the digital world enables working together while separated in place and time. We may no longer need the clock, with its rigid mechanical synchronization expressed in regular, uniform working hours. But trust, reliability, and collaboration—enabled by digital connections—remain central to industrial work.

THE GREAT TOILET PAPER CRISIS

In 2020, as an initial response to the onset of the COVID-19 pandemic, Americans stocked up on toilet paper. That impulse reflected an intuitive sense that the systems that support our lives are invisible when everything is stable and things work well but quickly fall behind with sudden changes in supply or demand. It turned out there was plenty of toilet paper. Producers stressed their systems to provide more without adding capacity, but the moment symbolized a larger realization.

Americans woke up to the fact that industrial systems support daily life, and those industrial systems are made up of people, machines, and infrastructure that are vulnerable to events.

Suddenly "supply chain," a term previously confined to academic researchers and logistics planners, was in the papers and on everyone's lips. The fraying sinews of global industry became visible to ordinary people. A new awareness arose that every grocery, every pair of socks, every electronic device is made somewhere and has to get from there to here.

White-collar workers stayed home and worked through screens, but people who worked in physical spaces moving the actual stuff around, also little noticed in public imagery, became suddenly apparent. These workers were dubbed "essential," even though the labor system they worked in treated them as dispensable and interchangeable.

The Ports of Long Beach and Los Angeles, which import 40 percent of America's containers, clogged to the point of gridlock. At one point, more than a hundred ships were sitting offshore at the Port of Los Angeles because no one was available to unload them when they docked. Construction materials became scarce. Essential workers—whether in factories, in hospitals, or at front desks in residential buildings—were most exposed to risk.

During the pandemic, people working from home consumed more goods, traveled less, ate out less than they did when commuting to offices. Urban air cleared. Lockdowns in China had as much impact on global supply chains as did work from home in the United States. Semiconductor shortages drove shortages of new automobiles; rising prices for cars led the growth in inflation. At one point in 2022, Ford had 40,000 F-150 trucks it could not ship because they were missing the blue oval Ford badges to stick on the front.

Heat waves, wind and snow storms, and other climate events stressed the problems further. When Russia invaded Ukraine in early 2022, another dimension became disturbingly apparent: these supply chains are heavily dependent on fossil fuels, and the longer they are, the more sensitive they are to the price of oil and gas, not to mention other global disruptions.

One of the first initiatives of the Biden administration in 2021 was to identify sources of supply chain vulnerability in semiconductors, advanced batteries, and pharmaceuticals. It found part of the problem in the decline of domestic manufacturing, especially in small and medium-sized enterprises, as a source of national vulnerability. Global dependence on Taiwan for advanced semiconductors, long understood by industry players but below the public radar, leapt to the fore as a policy issue.

SUPPLY CHAINS ARE US

Supply chains are large industrial systems. They are composed of heterogeneous elements, such as ships, aircraft, trains, and trucks, but also systems of labor, information, and finance that build them and connect them together. Usually the goods flow in one direction and money flows in the opposite direction. Their physical substrates are themselves industrial products, relying on ships, trucks, cranes, fossil fuels, and electric power, tied together by skilled human operators, supervisors, managers, and other industrial roles.

The linear suggestion of "chain" is misleading—a supply chain is a network or tree structure of suppliers supporting other suppliers, referred to in some industries as "tiers." In Intel's supply chain, for example, the mines that supply tantalum ore to go into chips are buried deep in Tier 12.

Each supply chain is a network of physical, contractual, financial, and social relationships that take resources and time to build. Companies often consider these networks among their crown jewels and closely guard their secrets. "Once one understands what is involved in just a single overseas shipment to a consumer's home," writes MIT supply chain guru Yossi Sheffi, "the question is not why the item does not make it on time but rather astonishment and wonder that such a thing can be completed in the first place."

In their heyday, department stores created dream worlds that you would walk through and peer into the edges of supply chains as windows to an apparently better life. Today the internet does a great job of abstracting away those systems, hiding this enormous complexity behind a simple friendly interface to replace the plush showrooms.

But your clicks have impacts in the human and physical worlds nonetheless. When you log on to an Amazon website, or any other e-commerce site for that matter, you are controlling a global supply chain. When you click "buy," you initiate a series of financial, information, and human events that result in a product appearing at your door.

Consider any product in your home. Where was it made? (That should be written on the label somewhere.) Where were the parts made? Who put them all together? How did it get to your doorstep?

Every object embodies answers to those questions, though it's overwhelming to think about every one. Nonetheless, it's a worthy exercise to do every once in a while for some objects, including your food. Throw out a moldy strawberry? Imagine the vast distances (and fossil fuels) that got it to your kitchen, only to miss being edible by a few hours. And of course the journey isn't over when you throw it in your trash. Nor is it over if you eat it; your digestion is part of the chain. Our act of consumption is just a brief moment in these ever-flowing systems.

And yet that moment brings with it some level of ethical responsibility for the systems that support it. How much do you want to know about the working conditions at every stage of the supply chain? How much are you morally obligated to know, with what consequences? Few of us would likely condone every moment of every supply chain for every product

we consume. Yet to fully ethically withdraw from that partic-
ipation would involve a renunciation of modern life. To para-
phrase the cartoon character Pogo, we have seen the supply
chains and they are us.

The pandemic era taught us that *disruption*, the Silicon Valley term that had been prized as an emblem of bold innovation, is not acceptable when it comes to supply chains. These
industrial systems are prosaic by high-tech standards—trucks,
workers in brown suits, and loading docks—but essential for
supporting lives in a modern industrial society.

Their goals, and the goals of the thousands of people who
plan, manage, and operate them, is the opposite of disruption.
Keep the goods flowing. Keep the power on. Keep the trains
running. Keep everyone safe while doing it. Smoothness, reliability, and efficiency are the watchwords. Dynamism is inevitable, but disruption is the enemy.

JEFFERSON'S ENLIGHTENMENT TEACHER

In 1765, as the Lunar Circle was forming, a young doctor appeared in Birmingham carrying letters of introduction from Benjamin Franklin. This earnest, engaging man, William Small, was arriving from a teaching stint in the new world. He remains the Lunar Circle figure least known to history, the only Lunar Society member who neither published scientific papers nor joined scientific societies. A biography of him appeared only in 2021.

Small preferred bringing people together: playing pivotal roles in sharing Enlightenment ideas and spreading them to a broader world. In fact, Small's untimely death would spur Boulton to establish the Lunar Society to compensate for the loss of his gregarious connecting friend.

Small was raised in Scotland and educated in the Scottish Enlightenment. Scotland had a special purchase on the intellectual currents of the day, as Scottish thinkers synthesized the work of Isaac Newton and Francis Bacon into the empirical, progressive application of knowledge. They emphasized numbers and counting, particularly as applied to political and social questions—including the measurement and maximization of happiness.

After joining Great Britain in 1707, Scotland had by the mid-eighteenth-century surged ahead in Enlightenment creativity. The *Encyclopaedia Britannica*, for example, the British counterpart to Denis Diderot's *Encyclopédie*, the ultimate Enlightenment project, was first published in 1768 in Edinburgh and has been continuously updated ever since. Other members of the Scottish Enlightenment included David Hume, James Boswell, Adam Smith, and Lunar Society members James Keir and James Hutton.

Small left Scotland, studied medicine in London, then traveled to Virginia in 1758 to the College of William and Mary in Williamsburg (itself founded by a Scot and built partly on a Scottish structure). When Small arrived, a contentious political battle (and much alcoholism) had led to the firing of the entire faculty. An inauspicious start for a bright young teacher: the colonies were a backwater, a distant, low station.

For a couple of years Small served as the only professor at the college, teaching every one of its classes. Through his teaching style, inherited from his Scottish mentors, he began to shift the college from a seminary-based theological education toward one focused on Enlightenment values of discussion and demonstration, science and literature. Small taught mathematics, chemistry, history, geography, ethics, and law.

It so happened that one of the students was a callow young slave owner who had come to party and drink. He settled down after a while, and while in college Small was his only teacher. His name was Thomas Jefferson, and because of Small he emerged from William and Mary a child of the Enlightenment (which seemed only to conflict, rather than reform him, on the issue of slavery).

The two became friends, and Small brought Jefferson into contact with the Williamsburg elite. Small even created a short-lived club devoted to the application of science and knowledge to practical problems, a "Society for the encouragement of Arts and Manufacturing." George Washington became a member.

"It was my great good fortune," wrote Jefferson in his autobiography, "and what probably fixed the destinies of my life. . . . that [my professor was] Dr. William Small of Scotland." Jefferson continued, "From his conversation, I got my first views of the expansion of science, and of the system of things in which we are placed."

Small might have been a mere pedagogical detail in Jefferson's biography but for the next move in his short life. In one of the pivotal connections of our tale, Jefferson's teacher became the man who connected Matthew Boulton to James Watt.

In 1764 Ben Franklin visited Williamsburg and met Small. Soon thereafter, probably at Franklin's suggestion, Small returned to London. Ostensibly he went to buy scientific instruments for the college, which he purchased from some of the finest craftsmen, with whom Franklin also collaborated. These included an evacuated pendulum, mirrors, telescopes, weights and measures, and electrical apparatus (Jefferson later ordered from the same makers for his own use).

Small shipped the instruments back to the college, but he never returned to Virginia. In London, Franklin took Small under his wing and introduced him to the intellectual and social life of the city, including a visit to the Royal Society in early 1765. Small returned briefly to Scotland to formalize his medical degree. Then Franklin sent him to Birmingham with a letter of introduction to Matthew Boulton.

Boulton immediately saw the extraordinary talent in Small, and hired him as a physician and scientific assistant. Small moved to Birmingham. After he operated on Boulton's daughter, Boulton helped set him up private medical practice. He quickly became prominent in the community, acting as a town commissioner and helping to establish a hospital.

It was in this year that Boulton and Small began to meet for "philosophical feasts." For the next ten years, until his untimely death at age forty-one, Small would advise Boulton on his scientific endeavors and gather the Lunar Circle.

CONQUERING WITH POTTERY

In the world of the Lunar Society, heroes were warriors, great statesmen, and kings. By the early eighteenth century, one might find a scientist or two in the pantheon—particularly Isaac Newton, as a reflection of the century of scientific revolution.

But that artisans, mechanics, or engineers might rise to comparable prominence was new. It remains a hopeful Enlightenment idea: one becomes known not through birth or conquest or power but through collaboration and building.

"Do you really think," Josiah Wedgwood asked his partner Thomas Bentley, "that we might make a complete conquest of France?" The potter was referring not to fighting or force but to commerce, "My blood moves quicker, I feel my strength increases for this conquest," he continued, "*conquer France by pottery ware.*" Wedgwood admitted that the notion sounded "vulgar," but it excited him nonetheless—replacing the old battles of blood with commercial and technical competition.

In Britain, popularizing this attitude owes much to British writer Samuel Smiles, writing fifty years after Boulton and Watt. Smiles did three things relevant for our rethinking of industry. First, he seems to have been the first historian to recognize the existence of the Lunar Society. Second, he brought the habits of mind of the Industrious Revolution to popular culture. He actually coined the term *self help*, still today one of the largest sections in a bookstore. Smiles's book of the same name became a runaway bestseller in Britain as a guidebook for the poor to rise to gentlemanly status through "diligent self-culture, self-discipline and self control." Chapter 2 of *Self Help* praised James Watt and other great inventors. For Smiles, the paragons of his middle-class values were the engineers.

Smiles followed *Self Help* with *Lives of the Engineers*, the title echoing the medieval text *Lives of the Saints*. This book devoted a chapter to Boulton, Watt, and the Lunar Society. "Watt was one of the most industrious of men," Smiles wrote, "and the story of his life proves, what all experience confirms, that it is not the man of the greatest natural vigour and capacity who achieves the highest results, but he who employs his powers with the greatest industry and the most carefully disciplined skill—the skill that comes by labour, application, and experience. Many men in his time knew far more than Watt, but none laboured so assiduously as he did to turn all that he did know to useful practical purposes."

Smiles's third contribution is that he made middle-class heroes out of inventors, what others still condescendingly called "mechanicks." In stark contrast to heroes of physical strength or military victory, these were creative heroes of diligence, enlightenment, and industry—values enshrined in the modern disciplines and cultures of engineering. (Isambard Kingdom Brunel, one of the great Victorian engineers, worked in a more romantic mode, building "castles in the air." Smiles excluded him from *Lives of the Engineers*, possibly because of his grand visions and risky behavior.)

Franklin's autobiography helped filter these ideas through the American lens and create a narrative of national character. Though he wrote it during the late eighteenth century (in part to reclaim his Americanness after living in Europe for decades), his autobiography's rise to middle-class mythology came later, roughly contemporaneous with Smiles.

Smiles also published full-length biographies of the engineers Boulton and Watt, potter Wedgwood, locomotive inventors George and Robert Stephenson, and others. These survive as more than morality tales, for Smiles often had early access to family papers, interviewed participants (including the engineers' children), and quoted at length from primary sources, with footnotes. While not perfect history, the books remain remarkably good reading.

One might still dismiss Smiles's and Franklin's values as antiquated Victorian prudishness. But if we recognize them as elevating heroic characteristics of building and improvement, they remain valuable voices in the evolution of industrial habits of mind.

Before he became famous in the 1980s for his televised interviews, Joseph Campbell characterized the Western hero cycle in his 1949 study, *The Hero with a Thousand Faces*. Building on Carl Jung's ideas, Campbell broke down the hero's story into a series of trials and episodes, found across numerous cultures, and interpreted the archetype as a story of self-realization and individuation. Moments in the cycle included the call to adventure, the supernatural aid, the teacher figure, the trials, conquest, and atonement, and the journey home. (By contrast, the Homeric engineer Hephaestus, was disabled, suggesting the technical expert's unsuitability for heroic action: proto-nerds, perhaps.) George Lucas modeled the *Star Wars* stories on the Campbell heroic cycle; their continuing popularity confirms the cycle's power.

More recently, a feminist critique of Campbell points out that in his accounts women play mainly roles as mothers, protectors, and icons of fertility and death. Yet women in mythology have been more central, and more heroic, than Campbell recognized. On quests and missions of their own, mythological heroines display additional qualities such as curiosity, empathy, resilience, collaboration, and storytelling. They also cunningly deploy technical skills, such as weaving and mending, as did Penelope in the *Odyssey*.

Smiles's engineer-heroes, and the members of the Lunar Society, could be seen as hybrids of these archetypes. Engineer heroes were not physically strong, did not geographically explore, and did not dominate through conquest. Smiles's engineers were curious, resilient, dissenting, collaborative outsiders, subversives to the heroic myths that surrounded them. Steampunks indeed.

For our purposes, we need not categorize any of these traits as masculine nor feminine (and heroines are surely prevalent in today's industrial pantheon). Rather we can recognize a moment in early modernity when Western culture's dominant heroic narrative began to change. Though today, the mythologies of our current techno-magnates may have reverted to more traditional heroes of domination (due in part to their own behaviors), so a similar updating of those narratives is in order.

No one today argues to revive Victorian (or Georgian) work habits or management practice—exclusively male management, long hours, no weekends, and child labor, not to mention industrial accidents and chemical exposure. Yet how to amend and augment industrial behaviors to contemplate a new Industrial Enlightenment? What ought to be today's habits of mind?

"TURNING THE MILL BY FIRE"

In 1765, when the newcomer William Small appeared in Birmingham, the connections among Birmingham's budding industrial-scientific elite continued to grow, both socially and geographically. Their early industries, from canals to steam engines, mines, and fossils, were intimately connected to the earth. In the quest for improved mobility, physical connections across the landscape led to social connections across towns and industries. Canals, the new technology of mobility, became a common interest among the Lunar Circle members. Nearly all owned shares in various canal ventures, and those ventures crucially expanded the group.

Erasmus Darwin became interested in one of the earliest of these grand projects, the Trent and Mersey Canal, designed to connect the ports of Hull and Liverpool, clear across the British island just north of Birmingham. Through this project Darwin met Josiah Wedgwood, the other Lunar Society manufacturer who would rival Boulton's importance. Wedgwood, like Boulton, was on the cusp of a major expansion of his vision and business—in fact, he would model his factory on Boulton's Soho (even buying equipment from him).

The Darwin and Wedgwood families drew close. Erasmus Darwin became the Wedgwood family physician, and the two families educated their children together. Darwin's son Robert would marry Josiah Wedgwood's daughter Susannah, who would give birth to their common grandson Charles, the famous naturalist, who would marry another of Josiah's grandchildren, his cousin Emma Wedgwood.

In 1766, Boulton and his second wife, Ann, moved their home to the site of the new Soho manufactory. By now their industrial works was a wonder of powered machinery, with trip hammers, polishers, stampers, and other machines driven by waterwheels from the mill pond.

But Boulton could not control the earthly force of this water and always fretted about its unsteady power. The pond produced less power in summer and none at all in winter when it froze, idling expensive equipment and people and delaying deliveries to customers. Though normally forward-looking, Boulton initially opposed a canal in Birmingham because it might divert precious water power from his factory.

To compensate for the variable water power, Boulton attached a horse-powered mill to drive the manufactory. But feeding and running nearly a dozen horses to drive the mill proved both costly and troublesome. "The enormous expense of the horse-power," Boulton wrote, "put me on thinking of turning the mill by fire." Here he would draw on mining technology.

Miners had been using steam-powered pumps for decades to pump out water, keeping the earth at bay as they drilled, dug, and bored ever deeper. A Newcomen engine with a twenty-four-inch cylinder was the size of a building, and it could lift a ton and a half of water with about fourteen strokes per minute. Because it operated with steam pressure of just over atmospheric pressure, it could use an ordinary brewer's boiler and other standard equipment of low precision. Unlike the later Boulton and Watt machines, Newcomen engines made few demands on the manufacturing technology of their day. But they consumed so much fuel they could be used only where coal was cheap—that is, near coal mines.

In his notebooks, Darwin had been drawing steam-powered carriages and contraptions for years. Boulton thought to build an engine of his own, to pump water up into the pond to drive his waterwheel.

Boulton sketched a steam engine design and built a model. In February 1766, he sent it to Benjamin Franklin for his opinion, "as the thirsty season is approaching apace." Boulton asked Franklin about valves and the best positioning for a cold jet of water to condense steam—indicating the fairly detailed level at which he was already working. Franklin didn't have much to say, pleading distraction with "our American Affairs," and returned the model. Undeterred, Boulton continued to experiment and correspond with Darwin on his steam curiosity.

Soon after, when Small introduced Boulton to James Watt, Watt found in Boulton a prepared mind. The man had a commercial need for a large, steady power source, saw the limitations of the water pond, and had an inkling that steam might provide it. Perhaps most important, he had tried a little himself and gained some sense of the practical challenges (valve construction, condensation).

To help expand the limited views of American industry by neoliberal economists, MIT has commissioned a series of studies over the past forty years to address current problems in manufacturing. These are not typical faculty-initiated research but rather task forces from across the institute composed of engineers, industrially oriented economists, political scientists, and others. Departing from the typical abstracted social-science approaches of macroeconomics, they instead drew from on-the-ground field research in hundreds of actual factories, and synthesized practical policy recommendations.

The first of these studies, *Made in America* (1989), responded to the slow US productivity growth of the 1980s, particularly the rise of Japanese and German competition in automotive and related industries. (As a young engineer, I read it in 1989, finding it so exciting it drew me to MIT as a graduate student.) The book only mentions China three times in passing, out of several hundred pages. But it was clear that other countries had learned to do manufacturing better than the United States, and the book explained exactly why manufacturing computer chips is not the same as manufacturing potato chips. One effect of *Made in America* was the creation of MIT's Leaders for Manufacturing program (later renamed Leaders for Global Operations)—a conscious attempt to make studying manufacturing cool again for smart young technical talent.

Jeff Wilke, an early graduate of Leaders for Manufacturing, joined a young e-commerce provider on the West Coast. The company was having trouble scaling its distribution operations. Wilke, who had worked in the chemical industry, pointed out that Amazon was conceptualizing its operations all wrong. Rather than imagining its operations as warehousing and distribution, Amazon should imagine them as manufacturing, and manufacturing in a continuous stream like in the chemical industry. Wilke reorganized Amazon's network, shuttered a few distribution centers, reorganized the others, and renamed them "fulfillment centers." The name, with its hint of spiritual fulfillment, sticks today. These key improvements enabled Amazon to scale to its current proportions. After leaving Amazon, Wilke has focused on rebuilding American manufacturing.

Later MIT studies included *Making in America* (2013), led by political scientist Suzanne Berger, who had worked on *Made in America*. The study, published after the great financial crisis of 2008, addressed the decline of US manufacturing in the early twenty-first century. This work concluded that manufacturing is not like agriculture, with an ever-declining labor share, and that it matters broadly in the economy. It influenced the Obama administration's Advanced Manufacturing Partnership, including the Manufacturing USA institutes. This national network of more than a dozen entities focuses on advanced fabrics, robotics, biomanufacturing, power electronics, and other technologies critical to manufacturing, modeled in part on Germany's Fraunhofer institutes, which help bridge research and industry.

In 2018, even before the pandemic, public discourse grew increasingly anxious that robots were coming to take away jobs, maybe all of them. Conversations included various jobs' "susceptibility to automation," and anxiety peaked about a second machine age with a potentially jobless future.

MIT created its next task force to address the future of work. For years, my research had focused on human roles in autonomous systems, in undersea exploration, aviation, and spaceflight—extreme environments that provide glimpses of the future of automation and work. MIT's president asked me to co-chair the project along with labor economist David Autor, with policy expert Elisabeth Reynolds as executive director.

We organized a group that included Berger, as well as experts on additive manufacturing (3D printing), autonomous cars, urban technology, artificial intelligence, cognitive science, robotics, Chinese political economy, and the anthropology of work. The subject captured the attention of corporate America: interest was so strong from outside MIT that we assembled an industry advisory board that included CEOs of Ford Motor Company, PepsiCo, IBM, and leaders of the country's major community college systems and labor unions. Leaders for Manufacturing graduate Jeff Wilke, who was by then CEO of Amazon's online retail division, also joined.

In our early conversations, economists in the group were asking, "How do we know what technology will do to us?" I responded by pointing out the key lesson from the history of technology: technology is not something that happens to us; it is made by people, with particular goals, making particular decisions.

Especially at a place like MIT, but throughout the economy, it is in our power to shape technology through research, the things we build, the stories we tell, and especially what we teach our students, engineers who will build the future. As consumers we constantly make choices about technology through the things we buy and how we live our lives. Rather than predicting some technological future from the pages of an overhyped trade press book, or from enthusiastic inventors, researchers, and entrepreneurs, our task became to understand how technologies are being deployed in the world, and how to build institutions to support the human role in those systems.

Teams of researchers and graduate students fanned out around the country and the world to look at the future of work. We formally titled the project "The Work of the Future," because work does have a future, and a major task for universities and MIT is to help figure out what that future could be and how to achieve it (we also liked the abbreviation WtF). The task force published studies that examined subjects such as healthcare work and AI, equality in labor systems in Northern Europe, warehouse and trucking technology, and industrial robotics.

Early conclusions indicated that the problem facing us would be not too many robots but not enough. David Autor, a labor economist, pointed out that the demographics and other labor market factors pointed to a shortage of workers for years to come. No evidence is pointing us toward a jobless future. One of the few things you can say about the future is how many twenty-year-olds there will be in twenty years. There won't be enough.

Automation, and mechanization before that, have always embodied human knowledge and skill into machines—that's the point, almost a definition of technology. One view of that process sees machines as putting people out of work or deskilling craft-type jobs into factory operatives. Yet those craft jobs are shaped by the technologies of a prior generation (like the job of machinist, which did not exist in the time of the Lunar Society).

Doing the same work with fewer hours of labor is the definition of increasing productivity. The important question is who reaps the benefits of that increase: workers or management, and in what proportion—a source of conflict since the earliest days of industry. In the United States, from the end of World War II to around 1980, workers in general shared the benefits of rising productivity.

From the 1980s till now, the economy experienced a great divergence, as those at the top end of the work scale have enjoyed the fruits of productivity and new technologies. Those in the middle have been largely automated away by technology, and those with the lowest-paid jobs—often doing work that is difficult to automate—have suffered.

Amazing digital tools that have transformed our desktops have offered the least to food service workers, cleaners, janitors, landscapers, and home health aides. These jobs are difficult to automate due to their physical demands and human interactions, yet they are relatively easy to train and hence low paid. Robotics pioneer Rodney Brooks points out that no AI can do all the jobs of a home health aide today, and the sum of those jobs (which include moving frail humans and sensitively communicating with patients, doctors, and family) might constitute a new Turing test for artificial general intelligence (AGI).

The jobs listed by the United States Census Bureau in 2018 show how the economy shifted from middle paid production and clerical work to highly paid professional and managerial roles. More than 60 percent of the 2018 jobs did not exist in 1940. Our ancestors could not imagine jobs like airplane designer, computer application engineer, circuit designer, wind turbine technician, or cybersecurity analyst. Not to mention tattoo artist, mental-health counselor, hypnotherapist, and drama therapist (what Autor calls "wealth work," jobs to service and support a new, largely urban professional class). Think about what you and your friends do, and ask yourself if those jobs even existed in your parents' time, much less your grandparents': solar panel installer, ride-share driver, website builder, computer network technician.

In most eras, where people were put out of work, new industries and new work arose to employ larger numbers of people. Of course, those displaced may not actually get those new jobs, and the personal disruptions are real, painful, and long-lasting.

Still, seeing automation at a larger scale than the existing task level shows how a virtual cycle of rising productivity can provide resources to invest in people whose livelihoods are disrupted by technological change. "The central challenge ahead," we wrote, "indeed, the work of the future—is to advance labor market opportunity to meet, complement, and shape technological innovation."

The "Work of the Future" study also concluded that however new technologies develop, if deployed into the same institutions we have, which were designed in a twentieth-century industrial world, will reproduce twentieth (or nineteenth) century labor conditions. The issue is not job quantity but job quality. The major contribution of the study (and the book of the same name) was to highlight the numerous ways that educators, engineers, entrepreneurs, policy makers, workers, users, and consumers can shape the technologies and policies that shape our world. Current technologies of AI make these issue even more pressing.

The Work of the Future projects were well under way, indeed nearing conclusion, when the pandemic hit in 2020, upending the world of work like nothing else has in our lifetimes. After an initial spike of high unemployment in June of that year, the problem of labor scarcity came roaring back. Nearly every industry struggled to hire and retain workers.

Some workers have found new power in these constrained markets, while others are still stuck without support. Sensible automation of lower-level tasks will relieve some of these pressures, as will accelerating information technology deployment to improve efficiency in industrial sectors. Few of these are shocking or disruptive; rather they constitute multidecade transformations that are only partly under way.

JAMES WATT, FRAIL CRAFTSMAN

It must have seemed a curious gathering place at the University of Glasgow in Scotland in the late 1750s. A young craftsman set up his shop on campus—full of delicate tools and curious mechanical devices. Here he would make and repair instruments for scientific demonstrations in the classrooms, apparatus for professors' experiments, even an organ for a local church.

Something about James Watt, the young instrument maker, his intriguing workshop full of unusual tools and instruments, the joy of watching him work while engaging his active mind, attracted students and teachers interested in the new mechanical philosophy. The shop became a social gathering place on campus. Adam Smith regularly dropped in, as did Joseph Black, a professor of chemistry working on latent heat. One can imagine the conversation as Watt filed at his vise, shaped on his lathe, or printed precision scales on instruments he made.

John Robison came into the shop one day and was pleased to find the instrument maker engaging in philosophical conversation. Robison was only twenty years old but already a worldly and accomplished maritime surveyor. Before returning to sea, he pointed Watt's attention to the Newcomen work in steam engines, planting a seed. (Among Robison's naval adventures he would conduct trials of Harrison's clocks for measuring longitude.)

James Watt, nervous and frail but also curious, mechanical, and inventive, was born in 1736. His home was Greenock, Scotland, a busy port thriving on the sugar trade. Mathematics and mechanical skills ran through the family for generations, though the Watts had all struggled economically.

Watt's father, a mathematics teacher, also plied hands-on trades: shipwright, carpenter, builder. He was prolific in his shop, turning out furniture, maritime pulleys (blocks), pumps, and numerous other pieces to sell for shipboard use. The elder Watt also invested in shipping vessels. As a colonial merchant, he supplied the North American and West Indies trades in tobacco, sugar, and enslaved people.

It is possible that Watt's father (aided by James himself) imported some slaves into Scotland. Watt and his family's connections to the slave trade has recently been a subject of historical revision by his former employer, the University of Glasgow, and other scholars.

A half century after the inventor's death, Samuel Smiles did much to make a hero of James Watt. Despite his admiration, Smiles always emphasized Watt's physical frailty and even questioned his manhood: "Struggling as it were for life all through his childhood, he acquired an almost feminine delicacy and sensitiveness, which made him shrink from the rough play of robust children."

Held home from school because of frequent illness and bullying, Watt explored the world of objects, developing a unified focus for mechanical things. He acquired dexterity with tools, working in wood and metal. Emerging in Watt was a character based not on physical strength or courage but on manual skill combined with paper abstraction and expert drawing.

Watt engaged the material culture of the industry around him, repairing the numerous instruments needed to guide the ships coming into his home port. He also read deeply in natural philosophy, conducted chemical experiments, and built some early electrical machines. When he was eighteen, he went to Glasgow to train as a scientific instrument maker.

Glasgow was a node on the Atlantic triangle, a premier port for Virginia tobacco, Caribbean sugar, and, later, Southern cotton from America. Yet it had no instrument maker at the time. Watt briefly apprenticed there, but he soon set out for London.

There he found a position and began apprenticing in brass work, building a host of measuring instruments, working into the night to avoid pressgangs from the East India Company. In 1756 he purchased a set of tools and returned to Glasgow to set up his own shop.

John Anderson, a professor of natural philosophy at Glasgow, served as a bridge between the university and local industry. He visited local artisans, shared scientific results with them, and imbibed trade knowledge. Watt knew Anderson, as he had been in school with his brother, and spent evenings at his home in scientific discussions. Anderson hired Watt, enabling his small workshop on campus and a showroom for selling his instruments. Watt struggled here as well. To survive, he began making musical instruments, at least several pipe organs, with skill but without commercial success. Still, the pneumatics and valves of the organs prepared him to work with steam a few years later.

Watt married his cousin Peggy in 1764. Though uneducated, she worked in his instrument shop and ran the business when he was away.

At Glasgow, Joseph Black was conducting the experiments that would lead to his theory of latent heat. A core idea of the later thermodynamics, latent heat described how substances in a state of transformation like melting or freezing absorb or emit heat while remaining at constant temperature. Black worked both as a professor and as an industrial consultant, giving him a wide view of practical problems and chemical applications.

At Robison's suggestion, Watt began to investigate the steam engines then in use for pumping out mines. The college had made a working model of a Newcomen engine for Anderson's teaching, though it had been sent to London for repair. Watt read all he could on steam and did his own experiments. He worked through the early mechanical textbooks derived from Newton and made his own engine models based on those of Newcomen.

While it's tempting to see Watt's work as an application of Newton's mechanical philosophy, machinery was itself a foundation for the Scientific Revolution. Not least of the revolutionary breakthroughs was the idea that the universe could be modeled as a machine, most notably as a clock. "To follow the *clock metaphor* for nature through the culture of early modern Europe," writes historian of science Steven Shapin, "is to trace the main contours of the mechanical philosophy . . . central to the Scientific Revolution." Clocks, machines, and their cousins, automata, seem self-acting and intelligent, even though we know they are not conscious agents, a characteristic shared by both the Newtonian universe and steam engines (not to mention today's AI).

Also note the importance of the air pump, "the Scientific Revolution's most important fact making machine," in the hands of Robert Boyle and the Royal Society he helped found. Steam engines were like exotic, advanced air pumps, driving pistons, seals, and vacuums with the regularity of the clock.

When the college's model of the Newcomen engine arrived from London, Watt repaired it and set about playing with it. It did not work very well, so he crafted a new boiler. At this point his experiments taught him that a small amount of steam could heat six times its weight of cold water to boiling—an early insight into the power contained within steam. Black explained the phenomenon in terms of his idea of latent heat.

Watt continued experimenting and measuring, always measuring: temperatures, elasticity, evaporation, pressures, and so on, until he concluded that much steam was wasted with the repeated heating and cooling of the engine cylinder. By now his business was thriving: he moved to a new shop with a partner/investor, employing sixteen people.

One Sunday in 1765 Watt went for a walk on the town green in Glasgow (one of the world's first golf courses). "I was thinking upon the engine at the time," he later wrote, "and had gone as far as the herd's house, when the idea came into my mind that as steam was an elastic body it would rush into a vacuum, and if a communication were made between the cylinder and an exhausted vessel, it would rush into it, and might be there condensed without cooling the cylinder. . . . I had not walked further than the Golf house when the whole thing was arranged in my mind."

Although he didn't tell the story until fifty years later, this was the supposed eureka moment of Watt's most famous invention, the separate condenser. "This capital improvement flashed upon his mind at once," Black later wrote in religious language, "and filled him with rapture." Watt quickly fabricated a model using his excellent mechanical skills, proved the principle, and declared "the invention was complete." He soon told Robison, "I have now made an engine that shall not waste a particle of steam."

Watt was right that at this moment he possessed key elements of his engine. Yet it would be more than a decade before he delivered a working engine on this principle, several years more before he could do so reliably and repeatably, still more before business success. "I can think of nothing else but this Machine," he wrote in April of 1765.

A thousand practical problems remained, not least scaling the engine from the simple laboratory model. Watt had yet to realize that the separate condenser introduced additional complexity: valves, water pumps, spray nozzles, linkages, air pumps, packing, and seals.

Figure 2
Valve gear on Boulton and Watt lap engine. Photo by the author, London Science Museum.

He would need to learn how to surround the piston with a jacket of steam to keep it warm, to make the engine double-acting so steam pressed the piston in both directions, and to regulate the engine's power with a governor.

Yet Watt could draw on no skilled machinists, no precision metal-cutting tools, no engineering standards. Unlike the relatively low-pressure Newcomen engines (which did not require mechanical precision), Watt's engines needed a mechanical industry that did not yet exist.

Watt didn't even have a common graphical language to specify parts. A modern mechanical drawing (even on paper) is a sophisticated technology, with symbols, conventions, standards, and tolerances to uniquely specify a three-dimensional shape made with particular techniques and finishes. But those drawing techniques would not arise until after Watt's death. Instead he dealt with sketches and verbal descriptions—and a great deal of frustration when mechanics did not build the parts exactly as he wanted them, or when the machines would not work properly once assembled.

That said, Watt helped nudge mechanical drawing into the industrial world. He made beautiful drawings of steam engine details, growing more sophisticated by year. In Watt's drawing, "we can see emerging, for the first time," writes historian Francis Pugh, "the conventions and applications that have continued to characterize engineering drawing down to the present day." Different component sets and varied steam and water flows on the drawings were coded in different colors. Sometimes the same drawing would stand for different sizes of engines, with numerical parameters in the notes to specify size. Watt plotted the internal engine pressures on graphs inside the drawings of the cylinders.

The existing Newcomen engines were enormous contraptions, built by millwrights. Watt was an instrument maker, not a millwright, and was little schooled in what he called "the practice of mechanics in great." Successively larger models of his engines evoked further challenges: boring precise cylinders, creating pressure seals between the piston and cylinder, valves that would not leak. Today we would call this scaling up.

Not least of the challenges was finding workers who could execute at the scale of mills with the precision of the instrument maker. Glasgow's blacksmiths and tinners were not up to the task. Nor was Watt up to managing them—he did not relate well to workers, finding them beneath him and the relationships exhausting. "My principal hinderance in erecting engines," Watt wrote, "is always the smith-work."

An entire technical system, indeed a budding industry of machine tools, would have to arise to make even one of Watt's engines. Watt continued to work on these problems for several years, but he lacked means and capital, and therefore he lacked time.

Black had been loaning Watt funds for the experiments, and soon brought him to a former student, John Roebuck. Roebuck had become a successful and enterprising chemical manufacturer and mine owner, and he was struggling with pumping out water from his coal mines. Roebuck had put up a Newcomen engine at his mine but found it disappointing. Here was a customer with a problem. On Black's suggestion, Watt and Roebuck met to discuss the engine.

Watt's efforts with his engine had hurt his instrument business; he still had to earn a living to support a family, now with two young children. In 1766 his partner in the instrument business died, so Watt sold the business and began doing more traditional engineering—surveying in particular but also industrial chemistry. He purchased an interest in a pottery firm, learning more about the management of heat. He also worked with Newcomen engines, gaining valuable experience with the core technology and with customers' needs.

Watt returned to the experimental engine work in 1768. Roebuck financed it and encouraged him to file a patent, in exchange for two-thirds of the business. The following year Watt began building a working-scale engine at a cottage outside Roebuck's Kinneil House, about thirty miles away near Edinburgh.

At the time, real money was flowing into canals, and Watt needed real money. He took a job as a surveyor for the Monkland Canal. When he visited London to secure a license for the canal, it was denied. He returned home disgusted but, fatefully, not before stopping in Birmingham (and possibly learning a bit about doing business in London).

In Birmingham, William Small knew Roebuck and had learned of Watt's steam endeavors through him, a fellow Scot. Small and Darwin met Watt and showed him the Soho manufactory. Boulton was away at this moment, but Watt immediately realized this was a different world, with its water power, machinery, and skilled mechanics. He could build engines there. The ordinarily secretive Watt shared the details of his engine experiments with Small and Darwin.

At this point Watt became part of Lunar Circle, with Small acting as a kind of agent connecting him to the group and recruiting him to Birmingham. "My idea was that you should settle here," Small wrote to him, "I should not hesitate to employ any sum of money I can command on your scheme . . . provided you will live here." Thus began a correspondence with Watt that would last the rest of Small's life.

Regularly discouraged and gloomy, depressed, and at times even suicidal, Watt continued working on the nearly full-scale model at Kinneil. It began running in 1769, but it too disappointed. The cylinder worked poorly. Piston sealing remained a problem: the older engines sealed the piston with water, but Watt's cylinder remained always hot, so water would not do because it boiled. Watt tried packing with cork, rags, even dung, but nothing held up under temperature, pressure, and wear. He worked on variations of condensers, steam jackets, pistons and seals, oil pumps, loading valves. Generally problems were in construction, but Watt also was in a state of permanent conflict with his workers, whom he called "villainous."

"You cannot conceive, how mortified I am with this disappointment," he wrote to Small of the failures with his engine. "It is a damned thing for a man to have his all hanging by a single string."

The work, and the demands of being away from his family, left Watt discouraged, despondent. Once again he gave up the steam engine project and took on surveying jobs, at first of land boundaries, then for a new canal. "The quiet and secluded habits of his [Watt's] early life did not fit him for the out-door work of the engineer," Smiles wrote. "He was timid and reserved . . . had neither the roughness of tongue nor stiffness of back to enable him to deal with rude labour gangs." It would be more than four years before he returned to the engine. In the first decade of his work after the Glasgow green revelation, Watt spent less than a third of his total time working on the engine, due to other pressing economic needs.

AUTOMATION AND WORK

Up until about World War II, the term *mechanization* aptly described technological change in a variety of industries, from meatpacking to automotive assembly. Regular, interlocking mechanisms, derived from clocks, reliably repeated motions in an endless series. Machine tools themselves may be the best example of mechanization, as they could coordinate multiple motions under the direction of skilled human hands. In the 1950s, the term "automation" arose to describe an era when machines were envisioned to replace human functions, often incorporating some kind of feedback from the environment.

Automation built on a host of developments from World War II in electronics, computing, and automated weapons that augmented human abilities in extreme environments. The famed Norden bombsight, for example, was an analog computer that worked closely with a human bombardier and an automatic pilot to calculate the precise bomb release time. Radar, too, was a form of automated seeing that used transmitters, antennas, and novel displays to expand human vision to tens or hundreds of miles.

Digital computers (initially derived from radar pulse circuitry) were at first understood as automation of tedious human tasks of calculating numerical data like ballistics tables for guns. Creative experiments with computers, however, building on wartime control systems, showed that computing and software could create flexible machines that complemented human capabilities. The Apollo lunar landings, for example, were heavily computerized, but the software fostered a rich sharing of tasks between people and machines that enabled the spacecraft to augment the pilots' capabilities, with great success. By contrast, Soviet spacecraft, which used older analog electronics, enacted a rigid automation that left less room for astronaut input.

When similar human-machine symbiosis found its way to personal computers, it engendered an explosion of growth in knowledge work—by augmenting people and enabling them to work in entirely new ways. Think of what spreadsheets enabled, from accountants to data scientists. Early indications about generative AI, particularly in augmenting coding, suggest that it can enhance productivity through augmentation in similar ways.

Yet public debates over technology and labor in the United States, particularly in logistics and transportation, still view automation primarily as a source of labor replacement. Some see the introduction of technology to the workplace as inherently deskilling and job destroying. Indeed, a great deal depends on management practices and whether they frame technology as a means to reduce workers rather than enhance existing jobs.

New approaches to technology in the workplace can enhance human performance—shifting some dull, dirty, and dangerous tasks while leaving human workers to exercise their judgment, experience, dexterity, and social relationships. A solution is augmentation over automation: use the fruits of today's technological development, from precise mechanics to robotics and AI, to aid workers in doing their jobs better, more safely, and in ways that enhance their autonomy.

To some degree we can shape technological development to benefit workers, but technology is far from the only factor. Automation in airline cockpits helped bring the crew from three members to two, but it also changed the way pilots fly and enhanced safety (though not without potential pitfalls). Management practices, training, and implementation details that respect and enhance workers as skilled users and contributors to the work are at least as important.

"TO SETTLE A MANUFACTORY"

In 1768 Watt traveled to London to pursue a patent for his steam engine. Again he returned through Birmingham, and this time he met Matthew Boulton. Watt had already seen the Soho factory, but now he toured it at the side of its progenitor. He and Boulton immediately connected at a personal level. Watt stayed for two weeks. In these early conversations took shape the outlines of industrial steam power.

Watt's flagging confidence got a boost from such a knowledgeable potential customer recognizing the value of his engine. Boulton was already sufficiently prominent that his mere interest in the Watt machine would boost its fortunes. But Boulton actually saw a great deal more in Watt's engine than the inventor himself did. Boulton shared with Watt a larger vision for the power of steam in industry.

Watt told Boulton he had good control of "constructing several working fire engines on the common principle, as well as in trying experiments to verify the theory," but was challenged by "the executive part of this." Because his sponsor Roebuck could not help with management, Watt wrote, "the greatest part of it must devolve on me who am from my natural inactivity, want of health & resolution, incapable of it." By contrast, Boulton had not only reputation and access to capital but management experience and a talent for human relationships.

When Watt returned home, he recounted his experience with Boulton to Roebuck, which piqued the patron's interest. Watt then reported back to Boulton that an offer of a one-third interest in the engine would be forthcoming from Roebuck, in exchange for paying half of the expenses for development. Watt was clear on the remaining challenges he faced: sealing the piston without water, finding appropriate oils, achieving sudden and strong enough vacuum. Watt also said that he and Roebuck would be open to including William Small in a partnership.

Progress on Watt's experiments, the filing of the patents, and even Boulton's interest encouraged Roebuck. Sensing the presence of a valuable invention, he began to pressure Watt to show more results. "You are letting the most active part of your life insensibly glide away," Roebuck chided him. "A day, a moment, ought not to be lost."

Roebuck was not paying Watt's salary, but he urged Watt anyway not to pursue other work, heedless of the engineer's need to support his family. Indeed Watt shortly sent a plan to Roebuck to build a full-size engine at his Kinneil workshop. He filed a patent in early January 1769. Boulton and Small advised him on the submission, suggesting he protect the fundamental idea rather than the design specifics. "You are certainly not obliged," Small wrote, "to teach any blockhead in the nation to construct masterly engines."

Buoyed by these milestones, Roebuck offered Boulton a deal: license to build Watt engines in three counties surrounding Birmingham.

Boulton refused the offer. His letter explaining his thinking to Watt points the way toward their shared industrial future. For he was seeing farther than Roebuck, farther than Watt.

"The plan proposed to me," Boulton wrote in response to Roebuck's offer, "is so very different from that which I had conceived at the time I talked with you [Watt] upon the subject." Boulton candidly offered the two motives that excited him about their earlier conversation, social and commercial: "love of you, and love of a money-getting ingenious project." The two men had just met, but already Boulton was speaking of love—the bond that would drive their industrial collaboration.

Boulton realized the engine would require three things to bring it to fruition: finance, "very accurate workmanship," and "extensive correspondence." Capital, precision work, and social relationships—these were Boulton's stock in trade. Boulton actually ceased his own engine experiments at this point, because he realized Watt was more skilled and further ahead.

But Boulton went on with what he imagined:

> My idea was to settle a manufactory near to my own by the side of our canal where I would erect all the conveniences necessary for the completion of engines, and from which manufactory we would serve all the world with engines of all sizes. By these means and your assistance we could engage and instruct some excellent workmen (with more excellent tools than would be worth any man's while to procure for one single engine) could execute the invention 20 per cent cheaper than it would be otherwise executed, and with as great a difference in accuracy as there is between the blacksmith and the mathematical instrument maker.

Boulton was not only thinking of building engines, he was thinking about something even further out: building a factory to build engines.

In this letter Boulton shared key ideas for the steam engine factory: Site it near his canal, enabling mobility for components and products. Recruit and train a labor force who could build industrial equipment with the same precision that Watt built into his scientific instruments. Provide them the finest tools that would allow them to work, "with as great a difference in accuracy as there is between the blacksmith and the mechanical instrument maker." Scale up the engine building. It would take more than two decades, but this vision would be realized as the Soho Foundry.

Because Boulton was already a manufacturer, he recognized the problems Watt was facing and had a plan to address them. His estimate of 20 percent savings seems to reflect a genuine calculation based on experience.

"It would not be worth my while to make for three counties only," he concluded in rejecting Roebuck's offer, "but I find it very well worth my while to make for all the world."

Boulton's letter, indeed this very paragraph, could stand as the founding document of the Industrial Revolution. Steam power, Boulton realized, would be of much greater impact than pumping out mines.

It would be four more years before Watt and Boulton joined forces as partners, seven before they actually shipped a working engine, and decades before they shipped uniform products. But Boulton framed the relationship, with love.

Small, Boulton, and Roebuck did come to an agreement later in 1769, to purchase, after twelve months and further experiments, a one-third share of the engine, though not in full partnership.

As the 1770s began, Watt was keeping a roof over his head but distracted with other engineering. Another trial of the Kinneil engine failed. Watt's fortunes would darken considerably before turning around. The following years justified his gloomy outlook.

Watt's wife Peggy died in childbirth with their third child, losing the baby too. He was now a widower with two small children to care for. Soon thereafter Roebuck's finances collapsed, and his creditors took over. Watt sank into despair.

None of Roebuck's creditors were interested in the engine project, so Boulton was able to buy the assets. At this, the lowest point in Watt's life, he sent his children to live with relatives and moved to Birmingham. In 1773 he shipped the Kinneil engine to Soho. Boulton was aware the engine patent was well into its lifecycle, so he encouraged Watt to extend it—which he was able to do for twenty-five more years.

July of 1775, then, saw the beginning of Boulton and Watt, the partnership created for the duration of the patent. A fresh start for Watt, but he was alone in a new world.

Boulton and Watt saved each other, although Watt was reluctant and Boulton's businesses were not going well. "He [Boulton] pleaded, he flattered, he pressured, pushed and finally pulled the reluctant Watt," writes historian Robert Schofield, "kicking and protesting, across the threshold of mediocrity into prosperity."

James Boswell, biographer of Samuel Johnson, visited Boulton in 1776 and wrote to Johnson noting, "the vastness and the contrivance of some of the machinery. . . . I shall never forget Mr. Boulton's expression to me: 'I sell here, sir, what all the world desires to have: POWER.'"

One can imagine Boulton sweeping his arm across the busy workshop, recognizing, indeed creating, the links between powers mechanical and powers of commerce and culture.

It was a famous moment, and one of extreme confidence. For this was still the beginning of the partnership, when Boulton and Watt were struggling to deliver their first engine.

Here they succeeded: in March of 1776 that engine began working at the Bloomfield Colliery, in a public event covered by the press, "so singular and so powerful a Machine . . . the Workmanship of the Whole did not pass unnoticed, nor unadmired."

HOW YOU GET TO WORK

It is 3:00 a.m., on a New York City subway platform in the pouring rain. I'm not returning from a party or heading to an early flight at the airport. Rather, my colleagues and I are here to test new technology on a few miles of track, developed by a company I founded, in a pilot project with the New York transit authority to show how methods for navigating robots can also be applied to subway trains.

Standing on these platforms, one doesn't have to squint too much to feel the nineteenth century. The girders holding up the station have an Eiffel Tower look to them, old-fashioned rivets thick with generations of overlapping coats of paint. Switches are the heavy steel of Carnegie and Pittsburgh. Signals are controlled by clanky electromechanical relays made in the 1930s. Green and red signal lights are large incandescent bulbs from the glory days of General Electric. Troops riding these trains to the Brooklyn Navy Yard in World War II, or my parents riding to Coney Island when they were children, would have recognized this mechanical landscape.

The New york City subway runs twenty-four hours per day. It has no downtime. Rather, the Metropolitan Transit Authority (MTA) blocks off a few hours here and there in the middle of the night for testing. It is a big deal and takes months of planning and approvals. Every minute is valuable—during testing, this section of track is closed, and passengers have to get off the train, board a bus for a few blocks, and then get back on the train. This is not the kind of "disruptive" that Silicon Valley likes to tout with its new technology. Disruptive here is a pain in the ass for everyone.

At three a.m., the subway in Queens is far from abandoned. In fact, the trains are packed. A few of the riders are partiers making their way home from bars. But most are people making their way to or from work.

The three a.m. crowd conveys a truth: the main goal of the New York City subway is to get people from affordable, often low-income areas to higher-income areas where they can work good jobs. And to get them home again. Every day. On time. Safely and with dignity.

Everyone at the MTA is well aware of this simple mission, and it helps get them out of bed in the morning. Because of its unusual geography built around an island, New York as we know it could not exist without the subway, doing its job day in and day out.

This mission sounds simple. But the requirements for reliability exist under hugely varying conditions—flooding (increasingly frequent), ice on the tracks, mechanical breakdowns, changes in passenger loads and demand, upgrading a system that operates 24/7. Not to mention the ups and downs of New York city and state politics. Getting five million people to and from work, every day, on time, with safety and dignity, is an enormous sociotechnical systems problem.

Ironically, the year of those trials was 2019, peak of the venture capital investments in autonomous cars. Billions of dollars poured into promises of a driverless future, with only the vaguest understandings of safety, cost savings, certification, even business models (a utopian vision presently coming to earth by economics and technical difficulty). Yet here was a (fully electrified) transit system bringing five million people to and from work every day, and here were just a few of a handful of engineers from a small company, trying to use new digital technologies to make it work better. We were not alone, but our efforts were dwarfed by the investments in cars.

The core technologies in our subway products and autonomous vehicles are the same: computing, sensors, navigation, control, and data. Yet the United States does not see comparable investment in improvement to this fully electric, green, highly efficient, proven method of moving people in and out of cities. We have overemphasized novelty and underemphasized the maintenance, upgrading, and improvement of the technologies that run our world.

DEATH FORMS THE LUNAR SOCIETY

Despite the promise of a new life in partnership for Boulton and Watt, darkness returned. "The last scene is just closed; the curtain is fallen," a distraught Boulton wrote to Watt in February of 1775, just months after the move to Birmingham. Boulton was reporting the death of their close friend, and potential business partner, William Small. "I have this evening bid adieu to our once good and virtuous friend for ever and for ever."

Watt himself had barely recovered from the loss of his own wife a year and a half before. He wrote back with the cool, suffering experience of an engineer, "It is our duty as soon as possible to drive from our minds every idea that gives us pain, particularly in cases like this, where our grief can avail us nothing." He pleaded with Boulton to "immerse yourself in this sea of business." The exhortations could not hide Watt's own gloom.

William Small died at age forty-one, possibly from malaria he picked up in Virginia. His loss threatened the close-knit Lunar Circle, for his charisma brought and kept them together. Compounding the loss of Small, Lunar Circle member John Whitehurst departed Birmingham for London (though he would remain in touch with the group).

Despite, or perhaps because of, Small's death, the period 1775–1780 saw the Lunar Circle consolidate into the Lunar

Society. Without Small, these friends, each in a busy, successful stage of life, would need a more formal ritual to stay together. The common link was an interest in natural philosophy.

Also at this moment Josiah Wedgwood became more integrated into the Lunar Circle. He had already proposed a cooperative research organization among potters, a "Club of Potters," to share information on stones, granites, and colors for ceramics. His idea did not succeed, but in the Lunar Society he found the collaboration he craved.

The group met mostly on Sundays in Boulton's home in Soho. Indeed, Boulton was the dominant and driving force behind the dinners. The first reference to a meeting is within three months after Small's death. Historian Robert Schofield concludes this was no coincidence, that the more regular meetings were a compensation for Small's absence.

The earliest recorded meeting is May of 1775, a gathering of Boulton, Watt, Darwin, Keir, and Withering. The first meeting called the "Lunar Society" occurred at the end of that year. One study estimates 130 meetings between about 1775 and 1805 (nearly matching the duration of the Boulton and Watt partnership and the steam engine patent). Topics of dinner discussion included Darwin's speaking automaton, heat transfer, the composition of inks, experiments to determine the composition of water, decimal weights and measures, and the education of children.

Letters between the friends begin to mention "lunar" meetings and meetings around full moons. "What inventions, what wit, what rhetoric, metaphysical, mechanical, & pyrotechnical will be on the wing?" Darwin asked Boulton of an upcoming meeting. "Bandy'd like a shuttlecock from one to another of your troop of philosophers!"

Schofield has aptly likened the Lunar Society to a modern industrial research lab—bringing together diverse ideas in the sciences to bear on practical problems. But unlike today's establishments, the Lunar Society never had any formal standing or membership list. Despite the industrial interests around the table, the group's primary interest was to discuss natural philosophy more than industrial application. The industrial contributions emerged organically from the relationships and ideas that enriched the members' work.

Early disagreement appeared within the circle about the American Revolution, just then boiling over into violence. Lunar Society member Thomas Day began writing pamphlets against the cruelty of American slavery. Wedgwood supported the American cause. Boulton opposed the rebellion. The matter seems not to have affected the friendships. Politics was not their main concern (though political violence would eventually end the group).

Over several decades, Lunar Society members included the pioneering geologist James Hutton, who first observed that contemporary rocks might reveal the earth's history, and iron master John Wilkinson, an early pioneer of boring machines, machine tools, and iron bridges, who would bore the cylinders for the Boulton and Watt engines. Canal engineer James Brindley, who engineered the Trent and Mersey Canal, also attended, as did artist Joseph Wright, who would become known for his paintings of contemporary science and industry, to this day the most popular visual depictions of the Enlightenment.

For the members, one compensation for the loss of Small was Joseph Priestley's move to Birmingham in 1780 to take a minister's job. As Schofield writes, Priestley "is surely the best example of the eighteenth century, dissenting, lower middle-class scientist whose background, training, profession and personal preference gave a deliberate orientation away from London toward the north and midlands of England." Priestley (born 1733) was a Calvinist, educated in one of the non-Anglican dissenting academies, and taught at the dissenting Warrington Academy. There he met and married Mary Wilkinson, the daughter of an ironmonger and sister of John Wilkinson, also a student there.

While teaching at Warrington, Priestley began his scientific work, with Benjamin Franklin's encouragement, in 1765. He was elected to the Royal Society the following year. His *History and Present State of Electricity*, published in 1767, became fundamental in spreading the state of the art and establishing a common foundation for researchers. He also conducted and published original experiments, and soon turned his attention to chemistry.

Priestley's 1772 publication *Directions for Impregnating Water with Fixed Air* introduced the notions and benefits of carbonated water (Johann Schweppe commercialized the idea as a popular drink just a few years later). Priestley later isolated and identified nitrogen, oxygen, nitrous oxide, sulphur dioxide, and ammonia, among numerous other contributions ranging from education to theology.

Lunar Society members thought so highly of Priestley that they gathered funds to pay him to do research, recognizing the potential practical benefits of his work. Priestley publicly praised the value the Lunar Society brought to his thinking (he was the only member to acknowledge the society in his publications). Alongside his scientific work, Priestley continuously published in religion and politics and advocated for the abolition of slavery. A decade after Priestley joined the group, backlash to those publications would hasten the end of the Lunar Society.

In 1781, Priestley published a paper describing "a mere random experiment, made to entertain a few philosophical friends, who had formed themselves into a private society, of which they had done the honor to make me a member." The work he described was electrical sparking of a mixture of hydrogen and air inside a cylinder, then observing the moisture it generated. From observing this experiment, Watt deduced the composition of water as a two-to-one ratio of phlogiston and dephlogisticated air (to use the chemical terms in use at the time). Watt was not first to publish the idea, and the discovery in its modern form of two parts hydrogen to one part oxygen is credited to Antoine Lavoisier.

Though all male, the Lunar Society was socially, philosophically, and industrially forward-looking, strongly anti-slavery, and generally favoring the American experiment. "The excitement of science and manufacturing," writes Jenny Uglow, "went side by side with experiments in living." Boulton and Wedgwood both educated their daughters, an unusual practice for the time, though they did not train them to inherit the business as they did with their sons.

Historians have noted the emerging separate spheres of domesticity and professional life during this period, but the Lunar Society had yet to see that division. Both Boulton and Wedgwood lived at their factory sites. For guests, visiting included domestic hospitality as much as factory tours.

Both of Watt's wives worked in his business; he entrusted them with operations when he was away. His first wife Peggy's domestic role integrated with her duties managing his shop. Watt largely worked at home, which included space for drawing and a workshop for his inventions and experiments (the workshop, from a later period of Watt's life, is now on display at the Science Museum in London). Watt's extended absences husbanding the steam engines cost a dear emotional toll, and his relationships with his two youngest children (with Peggy), James Jr. and Margaret, never recovered (though James Jr. took over the business).

Watt's second wife, Ann, was also involved with his business ventures, writing correspondence in his absence, even managing workers in Soho. Their house in Birmingham mixed office and domestic space as well.

Lunar Society conversation included education and psychology. Member Richard Edgeworth encouraged his daughter, Maria, to read Smith's *Wealth of Nations* when it came out, and she worked with her father managing the family estate. They later collaborated on a book, *Practical Education* (1798), which she likely mostly wrote, that aimed to bring science to bear on education. Their prescriptions for practical experiments and demonstrations read like today's project-based learning.

Over the years, Maria corresponded with Lunar Society members, including Darwin and Watt. She would go on to write more than ten novels, whose popularity made her among the best-known British novelists of the early nineteenth century and better known than most Lunar men in their day. Within her novels are a series of trials, experiments, and demonstrations that readers have identified as a "a Lunar-like interdisciplinary commitment to experiment and practical observation," including a character named Dr. X, likely modeled on Darwin. Maria Edgeworth's literary work is credited with shaping the outlook of Jane Austen.

Lunar Society members' curiosities reinforced each other. The discovery and excavation of buried cities at Pompeii and

Herculaneum sparked a craze for classical themes that Wedgwood engaged in his pottery. Wedgwood's canal building unearthed volumes of interesting fossils; studying geology helped industry turn those fossils into fuels, clays, and glazes. Wedgwood bought (and presumably read) Priestley's books on electricity and chemistry. Priestley tested material samples for Wedgwood (Appalachian flint and basalt from Mount Vesuvius), and Wedgwood supplied laboratory equipment for him. Priestley had cited Darwin's work in his *History and Present State of Electricity*, and Darwin and Franklin corresponded about Priestley's work. Watt, Small, Boulton, Keir, and Boulton had a plan to mass-produce clocks. When Boulton was in Cornwall delivering and supervising steam engines, he collected fossils and sorted them at night to relax.

Lunar Society members kept commonplace books for their musings and sketches. Darwin's, begun in 1776, was perhaps typical: in it he sketched everything from laboratory experiments to philosophical ideas, notes on medical cases to sketches of inventions for an artificial mechanical bird and perpetual motion devices. He drew maps, machines to weigh people, and steam-powered carriages.

Lunar Society friends corresponded on geology, chemistry, heat, glass, and combustion—each with obvious relationships to mining, pottery making, and metal and chemical manufacturing. In 1780, member James Keir cofounded the Tipton Chemical Works, which became one of the largest plants in England, producing alkalies for soap and other products.

While the group often met on the full moon, the lunar designation was as much symbolic as practical. Britain was full of coffeehouses, associations, and conversation, but the Lunar Society was unique in its industrial connections.

Soon after entering into the engine partnership with Boulton, Watt briefly returned to Scotland. He saw old friends and updated them on his work. He continued technical and scientific conversations. But his main goal was to remarry.

His second wife, Ann MacGregor, whom he wed in 1776, was a woman of deeper intellectual interests than the departed Peggy. She came from a bleach making family and conducted experiments with her father in chemical techniques. She would collaborate with James on the technical and financial aspects of his business.

As always, social capital and financial capital intertwined. Before agreeing to a dowry, Ann's father asked to see Watt's partnership agreement with Boulton. The only trouble: it hadn't been written down yet. Watt sent diffident, pleading letters to Boulton asking for his approval. Boulton verified Watt's account of the partnership and wished him well. Watt, consumed with his new hopeful future, ran through details of design. "I did not sleep last night, my mind being absorbed by steam." While in Scotland, Watt took orders for several engines.

Meanwhile, while Watt was away Boulton was busy fielding inquiries, taking orders, and building the engine factory at Soho. "I have fixed my mind upon making from twelve to fifteen reciprocating and fifty rotative engines per annum. . . . [O]f all the toys and trinkets which we manufacture at Soho, none shall take the place of fire-engines in respect of my attention."

Despite the interest in engines for factories, and Boulton's enthusiasm for them, the partners initially focused on sales to copper mines in Cornwall, which did not have easy access to coal. Building the business required nearly constant presence there from one of the partners.

And first they had to make the engines work. The Kinneil engine had a cylinder shaped out of tin; the metal was flexible but imprecise. Soon they found a key component, thanks to John Wilkinson (born 1728), Priestley's brother-in-law. Wilkinson (also a nonconformist educated at Warrington) was first an ironmonger, having developed a method of smelting using coal. He became known for his numerous iron inventions, including the iron coffin in which he would be buried.

The Iron business led Wilkinson into cannon manufacture, which required boring holes with precision in large cylinders. He developed a rigid boring machine that would create holes both straight and circular in the heavy materials. When he bored a new iron cylinder for the Kinneil engine, it proved much more satisfactory than the tin contraption. Boulton immediately ordered more; Wilkinson would bore all of the cylinders for the early engines.

Boulton wrote in 1776 that "Mr. Wilkinson has bored us several cylinders almost without error; that of 50 inches diameter, which we have put up at Tipton, does not err on the thickness of an old shilling in any part." The quote is revealing: Boulton was pleased with the minimal error in the boring, but his measure is "the thickness of an old shilling." Rather than measure tolerance in inches or thousandths of inches, as would soon become the norm, Boulton used an old coin. Nonetheless, the precision cylinder was the key to keeping the seal, the key to efficient, reliable operation, the key to the entire steam engine business.

Wilkinson then placed the first order for an engine from Soho, to drive the bellows of his ironworks. Watt delivered the engine parts and supervised construction. It worked well, impressing not only Wilkinson but neighboring ironworks, some of which had suspended construction on Newcomen engines, pending the new technology. Later Wilkinson also received the first rotary machine, delivered in 1783.

Boulton and Watt did not actually make engines in their factory but rather supplied drawings, fabricated crucial parts, and supervised the construction and operation of the engines at the customers' sites. In the early years, before setting up their own factory, they outsourced the parts fabrication to Wilkinson.

Much of the construction fell to millwrights, "a kind of jack-of-all-trades, who could with equal facility work at the lathe, the anvil, or the carpenter's bench. . . . He could handle the axe, the hammer, and the plane with equal precision. . . . he generally had a good knowledge of arithmetic, geometry, and theoretical as well as practical mechanics." When the parts arrived from different places, imperfectly specified and fabricated, the millwrights had to make them fit. Millwrights had the skills that Watt lacked, working machinery "in great," another source of the emergence of an engineering profession.

Boulton and Watt licensed their patent to customers for payment based on how much coal was saved. In today's lingo you would call this business model "steam as a service." It proved a never-ending source of headache and conflict. The mine owners resented the payments and threatened to turn to other steam engines. Watt constantly suspected, correctly, that the miners were cheating on their payments.

Watt invented a counter device to measure how much the engines were pumping—an early mechanical form of digital surveillance. Often Boulton and Watt had to take a financial interest in the mines in lieu of payment, drawing them unwillingly into the mining business (an expression of Boulton's ceaseless expansion that cost him focus).

The later engines used a less extractive fee scheme, charging royalties based on installed horsepower, rather than fuel saved. In fact, this strategy drove Watt to first quantify the horsepower in the early 1780s.

There were difficulties not only with machine tools but with workers themselves. "With our tools and our workmen," Watt wrote, "very little could be done." Watt found constant trouble with workmen in the shops and with the skills of those who built and operated the engines. The Wilkinson boring machine, and its successor machine tools, helped solve this problem—building skill and expertise into the machines themselves rather than relying on the craft skills of individuals. Though undoubtedly much skill was required to set up and operate them.

Watt moved to Cornwall with his family to supervise construction. It was a difficult life in an early industrial landscape: the rocky, wind-exposed peninsula in Southwest England, pocked by mines, equipment, and engines. "The spot we are at is the most disagreeable in the whole country," Mrs. Watt wrote to Mrs. Boulton. "The face of the earth is broken up in ten thousand heaps of rubbish, and there is scarce a tree to be seen."

To ensure success, in 1777 Watt himself completed and operated the first pumping engine at Chacewater in Cornwall. When it began running, it worked beautifully, pumping the mine. "We have had many spectators," he noted, "and several have already become converts." Both workers and mine owners came to see, and new orders came in. "All the world are agape," he wrote, "to see what it can do." The engine lowered the water level in the mine, where two earlier Newcomen engines had failed. Taking their place in ancient landscapes of faeries, pixies, and sprites, the chthonic steam engines mediated between human worlds and the earth below.

Despite the difficulty, Watt had a rare moment of hope. "By attending to the business of this country [Cornwall] alone, we may at least live comfortably; for I cannot suppose that less than twelve engines will be wanted in two or three years." In actuality, the market was limited, "but after that very few more, as these will be sufficient to get ore enough." Boulton agreed: "There is no other Cornwall to be found."

By 1778 Boulton was in Cornwall as well. His better business sense and human skills, not to mention a brighter outlook, became essential in maintaining relationships with the miners. He spent his days traveling around Cornwall visiting mines and supervising the erection of the engines. In the evenings, he would arrange the fossils he collected, sending them to Lunar friends like Wedgwood.

As a partner, Watt was a downer and an emotional drain. "Though we have, in general, succeeded in our undertakings," he wrote to Joseph Black, "the struggles we have had with natural difficulties, and with the ignorance, prejudices, and villains of mankind, have been very great." Watt never ceased complaining about the workers, the build quality, the amount of work. "Nothing but slovenliness," he wrote of the workers, "if not malice, is to be expected of them." He did not stop to think whether anything in his own behavior and expectations might have created the situation. Boulton realized he could not simply fire all of his workforce, as Watt urged him to do.

Watt's wife Ann despaired and took the unusual step of writing Boulton directly for help, out of concern for her husband's safety. "It is all that I can do to keep him from sinking under the fatal depression," she wrote of her husband. "Whether the badness of his health is owing to the lowness of his spirits, or the lowness of his spirits to his bad health, I cannot tell."

Boulton, for his part, was sunnier by disposition, despite severe financial pressures on his other businesses and the stresses and uncertainties of starting the engine business. While Boulton was patient with Watt and seems genuinely to have acted out of his original love for him, the difference could have threatened the business.

Constant complaints of idle, drunk workmen resulting in engine damage dogged the enterprise. Gradually, though, trusted and skilled assistants emerged. Exemplary of these was William Murdock, a Scot who joined Boulton and Watt in 1777 and quickly rose to prominence as one who could lead an entire construction crew in the field, impressing not only his underlings but customers as well. Murdock would achieve renown as an inventor in his own right.

During this time, Watt's fertile mind produced a host of other inventions. A letterpress machine helped copy the firm's business correspondence (one reason it is so well documented); its manufacture also proved profitable. Watt also created a screw-based micrometer to measure parts with fine precision, of a type still used in machine shops today.

Despite the grueling challenges, the royalties began to flow, in the thousands and then tens of thousands of pounds per year. By 1785, Bouton and Watt could pay off debts and experience some level of financial comfort for the first time in their lives.

Boulton, recognizing early the limits of the mining industry, pushed Watt to adapt the engine to industrial uses, as had been his plan from the beginning. "You seem to be fearful that mills will not answer. . . . I think that mills, though trifles in comparison with Cornish engines, present a field that is boundless, and that will be more permanent than these transient mines." Smaller engines made with more uniformity would help address the production problems. To do that, he needed new mechanisms from Watt.

At that time, steam engines used reciprocating—up and down—motion to raise and lower the pump. After visiting a copper rolling mill in Wales, Boulton realized that the key to steam engines driving industrial mills, as opposed to mines, would be rotary motion. After surveying customers (and perhaps selling to them a bit), "the People in London, Manchester, Birmingham," Boulton found, "are steam mill mad."

Some steam engineers actually thought it was impossible to generate rotary motion from steam. But Boulton made a working model, proving them wrong. He then pushed Watt to take up the task "of producing rotative motion from the vibrating or reciprocating motion of the fire-engine."

Rotary motion, Boulton realized, would make not just a better pumping engine nor even a better waterwheel. It would be a new machine that could decouple a mill from the need for siting near a waterway. A machine that could run during summer and winter, flood or drought. An industrial engine that merged the regularity of the clock with the power of fire, driving a mechanical industry. Indeed, steam-powered mills would come to run two clocks side by side, one of a traditional pendulum type ("clock time"), the other linked to the mill's machinery ("mill time"), showing close synchrony.

Yet twenty years after Watt came up with the separate condenser idea, he was exhausted. Every new customer, much less every new invention and new industry, became a burden on his time, strength, and character.

Despite Watt's protests, what followed became the most productive period in his career. He brought forth several major inventions, and numerous smaller improvements, to make the steam engine an industrial product. First, Watt applied a "sun and planet gear," invented by Murdock, to generate rotary motion. A simple crank might have done the job, but it had been patented (probably stolen from Watt) shortly before. Watt, dependent on his own patents for the Cornish engine, did not want to challenge another's patent, so he worked around it. (The sun and planet gear, while beautiful to watch, was replaced when the crank patent expired.)

Watt also made the engine double-acting, where steam pushed the piston in both directions. To achieve this, he added an elegant mechanism, known as a parallel motion, to replace a flexible chain and enable the piston to both push and pull the piston from the beam connected to the flywheel.

Watt created a mechanism to regulate the engine's speed—an idea he observed in mills, where a similar device regulated the space between millstones. The flyball governor, what he initially called a "Whirling Regulator," with its rotating spheres, kept the engine speed constant, the earliest industrial feedback mechanism.

A shaft from the engine spun a mechanism with two solid metal balls. As they whizzed around and as they sped up, centrifugal force lifted them outward. That motion linked to a steam valve, reducing the steam admitted to the engine and

Figure 3
Watt's parallel motion in action. Photo by the author, Manchester
Science and Industry Museum.

hence slowing it down. This negative feedback (faster spin-
ning, slower engine) had the effect of regulating the motion of
the engine in a kind of balance.

The governor became an eye-catching icon of industrial
mechanical motion. More important, the governor enabled
the engine to keep constant speed, regardless of fluctuating
steam pressure, loads on the engine, or variations in the mechanism or the environment. The governor effectively made the
steam engine into a clock, driving regular motion into the large
machinery of a factory. The combination of double-acting pistons and this regulation proved the key breakthrough to make
the engine attractive to the textile industry, which needed
smooth motion to avoid breaking fragile threads.

Successive patents in 1781–1784 included a number
of these improvements (though Watt never patented the

governor). Each of these, and numerous lesser improvements in valves, nozzles, and the like, even including one for reducing smoke pollution, entailed increases in complexity and cost, and depended on other work in fabrication, assembly, drawing, and uniformity.

And workers: the need for industrial work habits at Soho increased. In response, Boulton created a training program. The firm gradually found successful technical managers like Murdock who earned Watt's respect and could manage the workers in Cornwall and elsewhere. Over time, the general quality of the workforce increased with the rise of associated industries and the supply of skilled workers.

A strike of the engine smiths in 1791 over piece rates suggests the factory was not free of labor tensions. Nonetheless, in the words of one study, Soho became "a nursery for mill-wrights and mechanical engineers." Once again, the division of labor proved pivotal: there were millwrights, erectors, enginemen, pattern makers, nozzle fitters, and plumbers. Watt mostly produced drawings from his home while Boulton managed the works.

Gradually, the rotative engine became a self-contained industrial product. Mill owners had less technical expertise and equipment on hand than mine owners did, so they wanted a more unified machine. Boulton sought to "methodize the rotative engines so as to get on with them at a great pace." Boulton realized that even for the mining engines, standardized parts would make repair and replacement easier. "We are systematizing the business of engine making," Boulton wrote, "as we have done before in the button manufactory."

Early on, very little of the engine was actually made at Soho, except nozzles and valves. Parts were contracted to

Wilkinson and other nearby foundries and shops, or made locally where the engine would be erected. (Soho's relationship with Wilkinson soured after he was found to be selling unlicensed parts.) Gradually, to control production, the Soho operation grew, until it became its own facility in 1795, the Soho Foundry, built about a mile away to be closer to the Birmingham Canal.

By 1778, engine production had been standardized enough that Boulton and Watt embarked on another improvement in the process: they published a user manual, "Directions for Erecting and Working the Newly-Invented Steam Engines. By Boulton and Watt." Later a similar set of instructions was prepared for the rotative engine, in a single-page format intended to be mounted on an engine house wall.

"Everything must be kept as clean as possible," the operating instructions began, and "when the parts are oiled or greased, the old scurf must be taken off as much as possible." On one preserved copy, next to the last instruction on removing the dirty grease, a mischievous worker hand wrote, "The engine man to butter his bread with what comes off." Similarly, after details for packing the piston and cleaning the inside of the cylinder, the commentator helpfully added, "Don't leave your hat, coat, or shoes in the cylinder."

Starting a Boulton and Watt engine was a bit like playing a pipe organ (see figure 4). Moving elegantly curved valve handles, an operator introduced steam through an inlet into the cylinder to get the piston moving under human control. Flipping the valves exhausted the steam and introduced more in the other direction to reverse the piston.

After a few cycles of manually controlling accelerating speed, the operator flipped a lever and the valves mechanically engaged themselves. With cams connected to the overhead beam, the engine's motion took over its own control and became self-acting.

Sounds of clanking, pumping, flowing water, and hissing steam gave the whole machine a sense of vitality as the giant beam walked up and down and valves interacted like clockwork. Watt's engines even drew their own pressure curves on an indicator diagram, monitoring their own performance, as the governor balls spun to regulate their speed.

Boulton and Watt engines were literally spiritual machines, with sacred auras. An American visitor in Britain observed the shrine-like presentation of the engine in a mill, "An ornamental flight of steps facilitates the ascent to the machinery, and the stone floor is so nicely cleaned that one almost feels reluctant to sully it by the impression of dirty feet." And the silent reverence of watching the operation: "No rattling of the iron, or jarring sound is produced, and the only indication of your being where so great a power is generated . . . is the slight rushing sound of the steam, resembling the breathing of a slumbering infant." Subsequent observers of steam engines often note their life-like character. It would not have been unreasonable for the contemporary observer to describe the sight as evoking an artificial intelligence.

The first rotary engine was delivered in 1783 for a corn mill; the second was delivered to London to run the Whitbread brewery, earning a visit from the king. Soon orders poured in from London breweries. The product proved the key to success for Boulton and Watt. Wedgwood ordered an engine for grinding flints. Others ordered them for paper and cotton mills, flour and iron. In the coming decades, the company would receive more than a hundred orders for engines from abroad, including to the West Indies for driving sugar mills (places with particularly brutal conditions for enslaved labor).

Boulton installed an engine for lapping (polishing) his metal products at Soho in 1788; you can see it on display today at the Science Museum in London.

Within a few years Boulton was financially secure for the first time since founding Soho more than twenty years earlier. By 1800, there were 188 reciprocating and 308 rotative engines in operation.

Beginning in 1784, to show off their technology, Boulton and Watt built a complete mill, Albion Mills, in London as a showcase. King George and thousands of others, including Thomas Jefferson, visited the mill to see the new machines in action. It was the first engine in London with the parallel motion.

Watt was horrified by the public display, thinking it would give away his precious trade secrets. What Boulton saw as sales and public relations, Watt saw as ostentation. "Let us be content with *doing*," he wrote to Boulton.

The Albion mill burned down in 1791. Watt thought it arson—the mill had been vigorously opposed by local millers—though the proprietors insisted it was an accident. Despite the financial loss, it had served its marketing purpose and had only a minor impact on the booming engine business.

The poet William Blake lived nearby, and the husk of the burned factory may have been the source for his phrase "Dark, satanic mills," which came to stand for the Romantic critique of industrialism.

"GET EXCITED ABOUT MAINTENANCE"

While the United States has put much effort into building vast information processing networks, and the software that runs them, its roads and bridges, both literal and metaphorical, have fallen into disrepair. Even worse, they have become obsolete.

These systems are not mechanical, or autonomous, or purely technical—they are aggregates of people, infrastructure, and machines. And they require a great deal of human labor to keep them up and running in a daily sense. Maintenance is a relationship between people and things.

Parts fail. Machines break down. Systems mismatch. Most technical people spend most of their time doing maintenance. "Maintenance and repair are the most widespread forms of technical expertise," writes historian David Edgerton. In a survey Edgerton found that more than two-thirds of engineers are involved in maintenance. Yet this crucial work, "the realm of the small trader and skilled workers," has been overlooked by the public story of technology.

Automobiles, Edgerton points out, are made in a few large factories, but they are serviced in thousands of small repair shops—often by blue-collar workers working their way up a technical career ladder. Visit a place like Cuba, where automobiles from the 1950s still dominate the streets, and you'll see what concerted maintenance and repair can do with technologies long considered too old to run in the United States. As fleets of older cars shrink, they provide a supply of spare parts to those that remain operating. In developing countries, autos can reach a kind of equilibrium position where they can be repaired indefinitely.

An early motive for interchangeable parts was not necessarily production efficiency but rather maintenance: repairing guns on a battlefield was easier if any part could work with any gun (the same was true for Boulton's early effort at standardizing parts). Later, automobiles benefited from that standardization, as a cost of the complexity of piston engines is their intense requirement for maintenance.

Aircraft, too, require frequent inspections and maintenance—they are designed for repair and parts placement. While costly, aircraft can be kept airworthy for decades. We'd rarely consider buying a twenty- or thirty-year-old car, but a thirty-year-old aircraft can be upgraded (mostly its electronics) and perform perfectly well in today's world. The US Air Force has some airframes, like the B-52 bomber, well on their way toward a century of service.

A key element of improvement in industrial systems is not necessarily improving the performance but extending the time between maintenance cycles, and therefore reducing costs. Scholars Lee Vinsel and Andrew Russel have elevated the crucial role of maintenance—and the consequences of our financial and cultural underinvestment in this crucial human activity. "Older innovations," they write, "make up the fabric of our daily lives: electric power, reinforced concrete, the internal combustion engine."

They point out that much of what passes for innovation is actually "innovation speak," the breathless elevation of newness that by its nature devalues the work of most humans, "especially those who do the dirty work that keeps our technological civilization running." These are often called unskilled jobs, because those who classify jobs cannot see the hidden skills that workers develop in the course of doing their jobs. "Often talk of skills," they write, "is actually talk about social status and little more."

Russel and Vinsel have founded an online group they call the Maintainers, which calls for a renewed focus on maintenance and repair, the care work that tends to our systems. They argue that a maintenance mindset is crucial to any move toward technological sustainability.

In a similar vein, modern "right to repair" movements seek to enable easier access to electronic- and software-intensive consumer products, thereby reducing waste and costs to consumers. Part of this mindset requires recognizing the essential importance of things and the material world. The Maintainers call for a positive materialism that builds on the pleasures and social benefits from deep engagement in making, building, and repairing things.

Crucially, maintenance is not only a source of valuable work that conserves and protects the systems in place today. Maintenance breeds familiarity with the details of how machines and systems work, and therefore provides a site of expertise and improvement. As focused attention on the functioning of a system, and where it breaks down, maintenance reveals strengths and foibles on a day-to-day basis. Maintenance is process improvement that can lead to product improvement.

In 1776, the world changed, bringing inflection points in industrialization. That banner year saw the publication of Adam Smith's *The Wealth of Nations*, which laid down the principles of modern capitalism in opposition to the then-reigning philosophy of mercantilism. Edinburgh-based Smith was not himself a Lunar Society member, but a number of his friends were, including James Watt. It was also the year that Edward Gibbon published *The History of the Decline and Fall of the Roman Empire*, a cautionary tale for a growing imperial Britain. As if to symbolically offset Gibbon by sparking an industrial Britain, 1776 was also the year Boulton and Watt delivered their first steam engine.

And of course in 1776 the American colonies declared independence. One might call this shared birth year a coincidence, but it was a year that crystallized the effects of the Enlightenment in philosophy, economics, politics, and industry.

Lunar Society members indelibly linked the industrial revolution to the American revolution. William Small's teaching of Thomas Jefferson appears in the Declaration. Jefferson's influences have been difficult to trace, as a fire destroyed his personal library from early in his life. His intellectual genealogy was a matter of speculation among scholars who hunted for Jefferson's "lost world." Until the 1970s, scholars located the key roots for Jefferson's Declaration of Independence in the social contract of John Locke.

But historian Garry Wills found the key was Jefferson's obsession with numbers, his constant tinkering and inventing. "The real lost world of Thomas Jefferson," wrote Wills, "was the world of William Small, the invigorating realm of the Scottish Enlightenment at its zenith." Jefferson invented his estate of Monticello with its curious mechanical contrivances. He invented the University of Virginia. And, Wills, argues he invented the United States itself: "The nation is not . . . simply a found object, but a contrived thing, a product of the mind."

The first sentence of Jefferson's Declaration reflects Small's teaching.

> When in the Course of human events, it becomes necessary for one people to dissolve the political bands which have connected them with another, and to assume among the powers of the earth, the separate and equal station to which the *Laws of Nature* and of Nature's God entitle them, a decent respect to the opinions of mankind requires that they should declare the causes which impel them to the separation.

Jefferson used the words "Laws of Nature," not the "natural law" more familiar to lawyers and philosophers. The "Laws of Nature," argues historian of science I. B. Cohen, could only have come from Newton's *Principia*, which Small taught to Jefferson.

Consider, too, the second sentence of the Declaration: "We hold these truths to be self-evident, that all men are created equal, that they are endowed by their Creator with certain unalienable Rights, that among these are Life, Liberty and the pursuit of Happiness." The phrase "we hold these truths to be self-evident" echoed the axiomatic reasoning of Newton's

Principia, which held truths self-evident only to those who had adopted an entirely novel worldview. The "pursuit of happiness" echoes the Scottish Enlightenment.

On the day the Declaration was passed by the Continental Congress, Jefferson recorded in his diary not political adulation. Ever the Enlightenment counter and recorder, he rather noted the purchase of a thermometer and seven pairs of women's gloves, some accounting, and a few temperature readings. Through that same empirical mindset, the Industrial Enlightenment shaped the formation of the United States.

CULTS OF NEWNESS AND THE CHALLENGE OF ADOPTION

Today's public stories of technology run on a cult of newness and invention, relying on assumptions of instant adoption—the obviously better technology always wins, enthusiastically adopted by lead users who want to stay modern.

Rhetorical mantras emanate from Silicon Valley: "Move fast and break things." "Software is eating the world." Catchy phrases are then captured by a technology press and social media that amplify them into the next big thing, guiding investors, young engineers and programmers, even policy makers.

I once co-wrote a committee study for the National Research Council on government support for computer science, to show how federal support created the internet. The report ended by assessing the next big thing that would transform industry, education, and entertainment: virtual reality (VR). That was twenty-six years ago, 1998, and VR still struggles for adoption. Web3, the metaverse, blockchain, Industry 4.0—all technology-centric themes of inevitability. Some became useful technologies in the world, but few changed the tectonics.

Founders of companies are visionaries: they see a future no one else can see, and set about convincing everyone else that future is inevitable, in their own economic interest. I've been one myself. Some founders are indeed visionaries, and some of them are indeed right. But the overall picture of technology that remains, the one that saturates public conversation and shapes decisions, is shallow, utopian, marketing-driven.

When the world fails to conform to their visions, tech visionaries sometimes blame us. Pundits talk about "barriers" to new technologies—like ease of use, cost, safety, or ethical challenges, as if these legitimate social concerns were somehow standing in the way, rather than key dimensions of human welfare.

But it is actually quite normal to reject most new technologies. Even when machines change the world, they usually don't work very well when they are new. It's logical to keep doing things the way you have always done them, because you're more interested in getting your job done, getting through your day, than in using a new device or technique. Novel devices and techniques have to be quite a bit better to entice you to spend the effort to make the switch.

That's even more true for industrial systems. When working with those systems, getting through the day means meeting production targets, making the trains run on time, ensuring airplanes land safely. If you're responsible for something day in and day out, anything that's new better work right out of the box, or the system (and the people who operate it) will fall behind. The New York City subway is an extreme but by no means unique example: it runs twenty-four hours per day, seven days per week, 365 days per year. Just fixing what's broken, much less upgrading or changing how the system operates, entails major interruption.

It is not surprising that people resist change. What should surprise us is that anyone ever changes at all, that anyone ever adopts something and new. We should be asking: What made them change? What convinced them? This is the challenge of adoption.

In response to the manufacturing and supply chain crises of the COVID era, in 2022 the Biden administration and Congress passed a series of bills aimed at revitalizing industry in the United States: the Bipartisan Infrastructure Law, the CHIPS and Science Act, and the Inflation Reduction Act (which despite its name is really a climate technology bill). Together, these laws marked the largest advance in decades in industrial strategy, a major investment in technologies to advance national goals.

The infrastructure bill authorizes $1.2 trillion in spending over the next ten years. The goals are to improve the nation's transportation networks and its core infrastructure. The majority of the spending goes to states, by formula, for roads and bridges (which means the largest shares are for Texas and California, the states with the most highways). But it also boosts funding for passenger and freight rail (an existing, low-carbon transportation system), airports, ports, and waterways. It also aims to improve infrastructure for electric vehicles, public transit buses, school buses, and passenger ferries and funding to improve safety in all modes of transportation (a key process innovation).

The IRA (passed with only Democrat votes) promises nearly $400 billion in federal funding over ten years, through a mix of mechanisms. The vast majority goes to a clean energy transition and a significant amount (nearly $50 billion) to transportation, agriculture, and water. Most IRA funding is in the form of tax credits, as well as loan programs for energy infrastructure, clean energy, and making electric vehicles. These tax credits are designed to catalyze and encourage private investment. While most of the tax credits target corporations, they also include consumer incentives to encourage purchases of new, and for the first time used, electric vehicles, and home improvement tax credits for upgrades and heat pumps. The IRA has provisions to encourage quality jobs in the clean energy industry and to strengthen the STEM workforce pipeline to work those systems.

One way to think about the IRA is a new kind of commitment to decarbonization. Going back to Henry David Thoreau, and even to Jefferson, one approach to American environmentalism has been an anti-industrialism, praising the pastoral and longing for a new age of pre-industrialism.

"For decades, when environmentalists imagined an America that had begun to solve climate change," writes Robinson Meyer for *The Atlantic*, "they pictured many of the trappings of post-industrial urban liberalism: bike trails, micro-grids, and organic farms . . . but the revelation of the IRA is that decarbonizing the United States may require *re*industrializing it."

Indeed, the IRA moves away from a regulatory approach and toward industry: it contains funds for decarbonization programs in heavy industry and a new emphasis on industrial processes for the big four (ammonia, steel, concrete, and plastics). These include the critical demonstration projects that prove viability for scaling up to industrial systems.

The first part of the CHIPS and Science Act was authorized in 2021, driven by pandemic supply chain concerns for semiconductor chips. The Russian invasion of Ukraine helped push the funding appropriations for the bill (and added the R&D science part) over the political threshold, as politicians realized the United States is one Ukraine away from losing our major semiconductor supplier in Taiwan, if China were to make a similar brash move.

The United States currently makes 12 percent of the world's semiconductor supply, as compared with 37 percent in the 1990s. Even 61 percent of Intel's chip manufacturing is offshore, and the company no longer holds the leading edge. In 2020, Apple switched from Intel to TSMC for its most advanced processors. US semiconductor startups have needed to look offshore for venture capital to bring their ideas to production scale. In engineering schools, enrollment in computer science has soared, while plummeting in electrical engineering.

The most cutting-edge chips are not the entire story, for thousands of ordinary chips populate our cars, printers, and dishwashers. China makes 65 percent of these prosaic chips. Much of the cost difference with US manufacturers is no longer about labor costs but about the relative size of government subsidies.

The CHIPS and Science Act commits almost $300 billion over ten years, including appropriations for $53 billion to support semiconductor manufacturing, R&D and workforce development, $24 billion in tax credits for chip production, and $74 billion in loan guarantees for production facilities.

As of this writing, the science part of the act is only an authorization (with actual appropriations funding still pending) for a new technology directorate at the National Science Foundation as well as for new programs to improve wireless technology and strengthen semiconductor supply chains.

These include shoring up the defense industrial base for semi-conductors, as well as processing equipment.

So far, the government demand signal has evoked the intended response from industry. Following on the legislation, Micron, Qualcomm, and GlobalFoundries committed to invest in facilities in the United States, following Intel, TSMC, and Samsung.

The numbers are impressive: the entire Apollo lunar program cost around $200 billion in current dollars. So the new commitments represent ten to twenty times that amount, though on a more diffuse set of goals than simply landing a human on the moon, and over a slightly longer time period.

Economists have also begun to realize how federal support shapes economic geography and how the bulk of federal R&D funding has gone to a few cities and the coasts. Numerous regional cities have the preconditions for developing technical ecosystems, were the government to make that a priority. The recent bills have indeed internalized these place-based strategies, aiming to seed clusters of innovation that can generate critical mass to feed on themselves.

"Manufacturing is at last being understood as the crossroads between national security and economic security and the three are increasingly seen as interdependent." The success of Operation Warp Speed in creating vaccines in response to the COVID-19 pandemic, which required little basic science but a great deal of scaling up, testing, manufacturing, and distribution, showed that government could still accomplish important things when on a focused mission.

Beyond the vast numbers, these bills represent a transformation of US government policy toward industry. Since World War II, federal agencies have largely focused on the research stages of development in universities, national labs, and other research centers. The new policies extend that support to later stages and process improvement: prototyping, testing, demonstration, product development, even market creation.

The government's reluctance to support technologies farther along in their development cycles stems from resistance to industrial policy—that is, anything that seemed to interfere in the free market's ability to allocate resources. This reluctance came in part from economists' mathematical bent, their inability to model what they could not quantify, and an incomplete understanding of the dynamics of technological change. Adam Smith's oversimplification of the pin factory has been alive and well.

Critics call this "picking winners and losers"—that the government chooses which technologies should succeed, edging out market choices. Any number of failures can be pointed out: years of slow progress in AI, or the failed Solyndra investment during the Obama administration.

Yet in that same era, federal funds rescued Tesla from bankruptcy, subsidized consumer purchases of its cars, supported its charging stations, and enabled battery technologies. Tesla demonstrated that EVs could be made for the masses and spurred major stodgy auto manufacturers to embrace EVs. The current flowering of AI depended on many decades of federal support, with slow progress and multiple winters.

The new resourceces are impressive, to be sure, and of sufficient breadth and magnitude that they could help change the course of US industry.

Some actually find this new industrial strategy insufficiently industrial. It doesn't strengthen existing manufacturing institutes. It doesn't support R&D agencies to expand research in manufacturing technologies. It doesn't support scaling up manufacturing for startups. Still more could be done to transform the full innovation system to include advanced manufacturing. "We are at a once-in-a-century moment, where we could fundamentally change the way we undertake production."

Undoubtedly stories will arise of waste and failed investments. The successes may well be longer term and subtler than the failures, as they were for earlier interventions. Much depends on the government's ability to deploy resources intelligently and flexibly and to steer those resources to the right places—no easy task for technologies that still retain a fair amount of uncertainty.

Job-training resources will fail to be effective, too, if people do not want the new jobs being created. What will it take for an automotive assembly worker (or an urban dog walker) to retrain as an electrician or a wind turbine maintainer? How different are the jobs in a mechanical manufacturing plant versus one that produces batteries? How can we encourage students to be as enthusiastic about manufacturing as they are about AI?

While these bills garnered votes from both parties, they do not have universal support, and their future implementation could be in doubt depending on political shifts.

To return to the original question: What are we working toward? What is a vision of industrial society that achieves sustainability goals while still providing for human needs with high efficiency and low cost? What behaviors will be required to bring those visions to fruition?

Government intervention and support alone cannot usher in a new industrial era. The legislation and the spending have laid tectonics—public signals of support, training programs, and targeted spending—much as government did with interchangeable parts, railroads, automobiles, electronics, and the internet. Much will depend on the execution—how those funds are deployed intelligently versus spread thin to feed existing bureaucracies. How to effectively support small and medium-sized manufacturers? Can these resources shepherd adoption of new production technologies among large firms? What will be left once the funds wind their way through government channels, established institutions, and special interests?

Government support can shape but cannot force a change in American technological culture. That must come from the ground up, from technologists and industrial workers themselves, and from students and young people who choose to enter those domains. To transform industry requires some transformation of American culture, away from a lopsided focus on software-only innovation toward encounters with the material world. Less ladder climbing and more building things.

"TO ASTONISH THE WORLD, ALL AT ONCE"

> I scarcely know, without a good deal of recollection, whether I am a landed gentlemen, an engineer, or a potter; for indeed I am all three, and many other characters by turns.
> —Josiah Wedgwood to Thomas Bentley

Though always foremost a potter, Josiah Wedgwood's (1730–1795) identity confusion arose from the breadth of his interests and vision: a man with persistent class aspiration, an engineer with a desire to shape the world, and a humble potter who transformed his own industry, the aesthetics of his age, and English manufacturing.

Like so many industrial innovators, Wedgwood lived on the borders between classes: of humble birth, a craftsman, Enlightenment intellectual, and aspiring self-made gentleman. "This combination of master and workman in one person," his daughter wrote of him, "undoubtedly contributed largely to his success."

Wedgwood's products cohered an English identity for generations of consumers. Kings and queens bought his tableware. Edward Gibbon wrote *The History of the Decline and Fall of the Roman Empire* while surrounded by Wedgwood busts of classical heroes. Jane Austen, ever sensitive to subtle markers of class and wealth, praised Wedgwood pottery in her letters and novels. Wedgwood seems to have achieved his goal "to astonish the world all at once, for I hate piddling, you know."

Wedgwood's home town of Burslem, one of a collection of pottery towns near Stoke-on-Trent, was a humble place at the start of the eighteenth century. Where Birmingham's technical culture revolved around metalworking, Burslem, about fifty miles to the north and blessed with an abundance of clay and coal, specialized in pottery. Poor roads led to modest workshop-homes with discarded potsherds piled outside. Potters retrieved their clay from the ground, leaving empty holes outside their homes that filled with standing water and sewage.

Pots traveled to market on the backs of horses or donkeys. When the animals fell on poor roads in bad weather (as they often did), they could smash an entire load. English pottery in these years was ordinary, of modest quality at the bottom of the market, behind porcelain from China, Delftware from the Netherlands, and the expensive Limoges from France. Josiah Wedgwood would change that reputation, and with it the profile of English manufacturing.

Wedgwood, born into a large family of Burslem potters, was the youngest of thirteen children. "I myself began at the lowest round of the ladder," he would later say. His father died when he was nine years old, and Josiah was soon apprenticed in pottery to his older brother, "to learn the Art, Mystery, Occupation or Imployment of Throwing and Handling." His brother put him to work making ordinary earthenware like butter pots, jugs, and washbasins. In an early division of labor, the young Wedgwood proved talented in modeling and throwing.

An epidemic derailed the promising craftsman. Smallpox swept through Burslem, striking and nearly killing young

Wedgwood. The infection left him with an agonizing pain in his right knee that would disable and plague him for the rest of his life. The stiffness of the joint meant that Wedgwood could only sit with his leg extended, which prevented long sessions at a wheel throwing pots (later he would have the leg amputated, while he watched, fully awake).

An injured workman was nothing new. But here the disability spurred energy and curiosity. In retrospect, the calamity "proved the turning-point of Wedgwood's career." Josiah's damaged body prevented him from taking the path of a traditional craftsman, so he turned his attention to contemplation, tinkering, and business. Thus began a lifelong experimentation with the processes of pottery—heats, kilns, glazes, and so much more, though always built upon manual skill working the clay.

Wedgwood's brother, and those around him, practiced a craft they had been taught for generations. His brother, his master, had no interest in precocious improvements, so when the term of the apprenticeship finished, Josiah left.

He soon joined the firm of Thomas Whieldon, an eminent potter and expert in the division of labor who recognized the young talent and supported his development. Wedgwood experimented with glazes. He soon introduced a new green with a hard, shiny finish, which made beautiful pottery and could even imitate jewels when mounted in metal. Wedgwood's new glazes captured green and yellow hues for the cauliflower and colonial pineapple shapes of the then-popular rococo style.

Shifting fashion and taste meant manufacturers had to be constantly alert to product and process innovation. Looking ahead, Wedgwood understood that the current stoneware (with a Burslem-developed salt glaze) was familiar to consumers, and hence declining in price, "and with regard to elegance of form, was an object very little attended to." One word constantly shows up in Wedgwood's notes: "improvement." "I saw the field was spacious," he wrote of the future of pottery, using an earthly analogy for the market, "and the soil so good as to promise recompense to any one who should labor diligently in its cultivation."

Wedgwood set out on his own in business in 1759, at age thirty. He worked every aspect of the pottery production process, making models, overseeing the firing in the kilns, and supervising workers. He continued to make green glazed ware, as well as tortoiseshell items and tea services.

As he began to achieve some success, he expanded to additional buildings in Burslem. Wedgwood became dissatisfied with the discipline of his workers, who showed up to the works at irregular times. At the center of one property he built a cupola with a large bell to call the workers to the site. It became known as the Bell Works, symbolizing an ongoing struggle between Wedgwood and his workforce. New methods, new products, required new, regular habits underlaid by discipline in time.

The early 1760s saw Wedgwood's personal and business status growing. He married up, to his cousin Sarah Wedgwood from a prosperous cheese-making family. In a long and happy marriage, she learned his techniques and helped operate the business. As in Boulton's marriage, her dowry brought capital to expand the company.

On a trip to Liverpool in 1762, Wedgwood fell off his horse and further injured his bad knee, requiring weeks of recovery in the city. The enforced stay away from home brought new friends through his surgeon Matthew Turner. Turner was about to begin lecturing to Joseph Priestley in chemistry, and he introduced Wedgwood to the promising pastor; their common chemical interest sparked an immediate friendship.

Another of these new connections, Thomas Bentley, became Wedgwood's agent in Liverpool. Bentley was a founder of the leading dissenting academy, Warrington, where Priestley taught. Bentley mentored Wedgwood, gave him books to read, and introduced him to scientific culture, as well as to the writings of Rousseau and Malthus (a Warrington student) and in person to Benjamin Franklin.

An experienced merchant in Liverpool, Bentley also provided the business experience, commercial network, and social introductions to propel Wedgwood from a local manufacturer to a figure of national importance. For years, the two corresponded multiple times per week, in affectionate, loving tones. Bentley had powerful friends, and he had taste. He would be Wedgwood's antenna to fashion and markets. He had the looks, the dress, and the manners to sell pottery to the upper classes. He would introduce Wedgwood to collectors who would enable Wedgwood to capitalize on the craze for classical themes. After moving to London, Bentley provided crucial feedback on taste and pricing.

Wedgwood had been experimenting for years, but soon after Wedgwood met Bentley and Priestley, his notebooks begin taking on a more systematic bent, less simply empirical and more informed by science. To hide his data from prying eyes, Wedgwood coded his materials and experiments in a cipher of numbers and letters.

The key to the cipher reveals a diligent application of empirical observation: "G.O. signifies the gloss oven; B.O. the biscuit oven; W.O. the white oven; and the letters B.M.T. prefixed to these mean the bottom, middle, and top of the respective ovens." Experiments ended with commentary:

"This merits further trial; try it again" or "The crucible broke, try it again." The writing reflected Enlightenment habits of mind: precision, recording, observation.

Pottery is an earthly technology, one of humankind's oldest crafts, transforming humble dirt into useful forms. The potter's skill depends on intimate knowledge of broad varieties of earths and clays. Wedgwood's artistic designs also depended on natural forms, which he sketched in his notebooks at home and while traveling. He read widely on mathematics and geography and mined travel books for clues to local minerals. When he borrowed texts on chemistry and clays, he would sometimes copy the works by hand for later reference. Wedgwood amassed a large collection of earths, stones, and clays from around the world, comprising more than seven thousand specimens.

In thousands of experiments, Wedgwood engaged core questions of the scientific Enlightenment: the generation and measurement of heat, the sources and transformations of materials, the distribution of those materials in the earth and their chemical combinations. Synthesizing them all through times, temperatures, and sequences would comprise the secret industrial recipes of Wedgwood's process.

In 1763, Wedgwood created a new form of durable and inexpensive tableware, which became known as creamware. It featured a novel combination of local clays, flint, and glass, fired in a complex new method Wedgwood invented. "When prepared in perfection," he wrote, "the ware may be considered as coated over with real flint glass." The product was so white, hard, and smooth that it could compete effectively with Chinese porcelain. It could also accept a wide array of glazes, paints, and designs in a hard, durable finish.

Wedgwood and Bentley recognized that taste as a social phenomenon flowed from the top. They marketed their products to noble customers—today we would call them "influencers"—beginning with tiles and tableware to decorate country manor houses. When dukes and duchesses used these items, the middle class wanted them more. Like Boulton, Wedgwood opened his factory to noble visitors, and it soon became a tourist spot.

But Wedgwood set his sights even higher. In 1761, the newly crowned King George III married the Hanoverian Princess Charlotte. One of the new queen's maids of honor was from the Staffordshire region that included Burslem, and she became aware of Wedgwood's rising reputation. Bentley suggested that she recommend to the queen an order for the new creamware. Wedgwood provided samples of his newest techniques, proposing "two sets of vases, cream-colored, engine-turned and printed." He received the order and took pains to execute it with the highest standard of quality and beauty.

Queen Charlotte was pleased with the initial pieces and soon ordered a complete set of table service. She suggested it be dubbed "Queen's Ware" and that Wedgwood be named "Potter to Her Majesty." The king soon followed with a separate order for what became the "Royal Pattern." Though a traditional form of royal patronage, the queen's purchase also represented state support of Wedgwood's technology at a crucial moment, propelling his domestic products ahead of high-quality Chinese imports. She also underscored what Wedgwood and Bentley intuited: female domestic taste and aspiration would drive this industry.

Wedgwood's customers consumed within a global network, for he rode the great wave of tea drinking. Today we associate Englishness with tea, but the leaves came from China. Tea only became English during the eighteenth century, during which English imports expanded by a factor of two hundred, thanks to the East India Company. An expression of empire, tea evolved from an aristocratic drink to a daily pleasure of middle classes and laborers.

The partner to tea was sugar, from enslaved workers on plantations in Barbados and Jamaica. Sugar consumption in England quadrupled during this same period, to more than twenty pounds per person per year. Some sugar enhanced tasty desserts, but most sweetened tea. Like today's economy, the feverish pace of the Industrious Revolution relied on copious supplies of sugar and caffeine.

Along with all this tea drinking came demand for "service." Pewter, hot to the touch and with a low melting point, is not suited to hot drinks, so consumers turned to ceramic pottery: teapots, cups, saucers, jugs, caddies, strainers, mugs, and sugar bowls. Thanks to Wedgwood and his industry, these items not only held the stuff but signaled Englishness, status, good taste, and modernity.

Wedgwood improved a technique to transfer printed images to the pottery, which worked particularly well with his creamware. Pictures expanded ceramics' potential for cultural variety and representation.

In 1773, Catherine the Great of Russia ordered a large service of almost a thousand pieces, each to portray a scene of England. Wedgwood invested massive effort to deliver the "Frog Service," named after the amphibian mascot of the Russian palace that would house it. Deploying Wedgwood's image transfer process, the pottery depicted manor houses, ancient ruins, and landscapes, encapsulating a picture of Englishness for the foreign royal customer. Among the scenes, Wedgwood included his own factory (with the Grand Trunk Canal in the background) and a few images of factories, mills, and bridges—a newly industrial, if still idealized, picture of England.

Wedgwood's pottery served as a media technology, conveying stories, culture, and national character. Most of Wedgwood's production went to export. The American colonies, with little pottery manufacture of their own and growing wealth and desire for English status, emerged as a major market. Wedgwood shipped a set of dinner service to George Washington at Mount Vernon.

Wedgwood, who was the same age as Matthew Boulton but less established in his career, admired his metalworking friend. "He is, I believe," Wedgwood said of him, "the first, or most complete manufacturer in England in metal. . . . very ingenious, Philosophical, and Agreeable." The two briefly became business partners. The partnership never took off, in part due to the differences in process between metalworking and pottery, but the friendship flowered and deepened through the Lunar Circle.

On a visit to Boulton in Soho in 1767, Wedgwood noticed a metalworking lathe that sat unused. He bought the machine from Soho to make novel patterns on pottery. Again, the theme: precision, this time in decoration. Cams programed regular motions in rotation to produce precise, geometric patterns. Wedgwood eventually designed his own pottery lathe with changeable cams to enable flexibility in design.

As with today's semiconductor chips, Wedgwood's key components came from elsewhere: flint, porcelain clays, unusual minerals. Each of these, and their finished products, traveled over bad roads from other parts of England or by sea via the port of Liverpool. Wedgwood recognized the importance of these networks and began advocating for improving the roads by connecting the potteries to newly emerging turnpikes.

Yet as Wedgwood's production capacity grew, he remained concerned about distribution. The nearest ports in Liverpool,

Bristol, and Hull required overland transits of eighteen to forty miles, which much of the product would not survive. Lobbying for turnpike improvements led him to the next technology: canals.

Wedgwood had known James Brindley before he became the nation's leading canal engineer. Brindley started as a millwright and built a flint mill near Burslem to supply the manufacture of whiteware. Later, Brindley vaulted to fame when he collaborated with the Duke of Bridgewater on Britain's first major canal between Manchester and Liverpool, which opened in 1761.

Yet Brindley had a vision for a broader network connecting multiple canals along a central trunk. He surveyed a route between the Trent and Mersey Rivers, which would help connect the ports of Liverpool and Hull across the British isle. Brindley and Wedgwood went to consult the duke; after they pitched the project during a canal ride in the duke's gondola, the duke immediately placed an order for a complete set of Wedgwood's cream-colored tableware.

Then as now, big projects required patronage, politics, and public relations. Wedgwood published a pamphlet to argue for the benefits of the canal. While promoting it, Wedgwood became acquainted with Erasmus Darwin, who began commenting on it to Wedgwood's partner, Bentley (insulting and annoying him in the process).

Darwin became enamored of canals, sketched inventions for locks, and supported the application. Darwin brought the canal idea to Boulton and Small, drawing Wedgwood into the Lunar Circle. "I hope you will give this Scheme your assistance," Darwin wrote. "I desire you and Dr. Small will take this invention, as you have given me ye Infection of Steam-enginry."

Wedgwood traveled to London with Bentley and Darwin to promote a bill for the project, which made its way through Parliament and was approved by the king in 1766. Wedgwood made the initial investment of a thousand pounds into what was officially called "The Canal from the Trent to the Mersey." But Brindley had a grander vision and renamed it "The Grand Trunk Canal" because he saw, correctly, the myriad tributaries it would spawn.

In 1766, it was a day of much celebrating in Burslem and the pottery villages when Wedgwood dug the first shovelful of sod for the canal. To accommodate Wedgwood's disability, Brindley carted it away in a wheelbarrow. The whole town celebrated well into the night. When finished, the Grand Trunk Canal would span almost 140 miles, with seventy locks and five tunnels.

Inspired by Boulton's Soho, Wedgwood immediately purchased the land for a new factory, right along the route where the new canal would run, extending to it with a short branch. Erasmus Darwin suggested the name "Etruria." The name would stick, and the Etruria factory would soon outshine Soho in the fame of its process and the attractiveness of its products.

"Etruria" was an ancient allusion. Despite its modern aspirations, British industrial culture in the eighteenth century drew inspiration from classical antiquity. In the 1750s, Enlightenment-inspired excavation at the foot of Mount Vesuvius near Naples began to reveal the wonders of the buried cities of Pompeii and Herculaneum. Architecture soon began to show off the "antique manner," better known as neoclassicism.

The British envoy in Naples, Sir William Hamilton, became enamored with antique artifacts, particularly the vases recovered from nearby Etruscan tombs. The Etruscans held fascination for the English as the enigmatic predecessors to the

Romans. Hamilton amassed a vast collection, which he later sold to the British Museum. He sponsored artists to sketch the vases, and then published them in a series of illustrated books, which brought the classical style to popular attention (some have called them the most important art publications of the eighteenth century). Hamilton, himself a customer of Boulton's metalwork, thought the catalog would be particularly useful to British "manufacturers of earthenware and China."

Wedgwood, ever alert to stylistic trends among his aristocratic patrons, eagerly dived into the Hamilton catalog. He and Bentley began copying the vases and updating their designs. "With their characteristic nimbleness and unsentimental realism," writes Wedgwood biographer Tristram Hunt, "Wedgwood and Bentley's business was ready to place the past at the very forefront of modern, industrializing Britain."

Wedgwood had several lines of ceramic ware in production, but he became so enamored with the Etruscan styles that he added black basalt vases to his line and decorated them with classical themes, imitating the ancient red-figure vases. Wedgwood again had to improve the process: the ancient techniques had been lost, and Wedgwood had to reconstruct and update them. The experiments were complex and prone to failure, but through a "meticulous, iterative approach" Wedgwood was able to "improve the Art and Mystery pretty fast." The only patent Wedgwood sought and was granted in his career was a method of copying classical vases into modern forms.

The name Etruria was actually a misnomer. Most of the vases Hamilton and Wedgwood so admired, though found in Etruscan tombs, were not Etruscan but were sold to them by the Athenian Greeks in the sixth to fourth centuries BC. The Etruscans then used the vases as burial objects (though some later Etruscan and South Italian potters copied them). Indeed, the Athenian workshops, turning out large numbers of beautiful, stylized products, resembled industrial production in the scale of their output, the division of labor, and the standardization of their vessels.

Nonetheless, the name Etruria stuck, and Wedgwood called his workers Etruscans. "Here we have a colony raised in a desert," wrote one visitor to the factory, noting the provincial outpost, "where clay-built Man subsists on clay, and where he seems to want nothing but the power of Prometheus to copy himself in that material."

British aristocrats began to decorate the mantelpieces in their neoclassical mansions with Wedgwood pottery. On the cusp of industrialism, Wedgwood projected a simpler time into the past and drew on its clean, simple, humanist lines to create a modern design. Demand soared: by 1765, Wedgwood was writing to Bentley that "an epidemical madness reigns for vases." Bentley moved to London and opened a showroom where fashionable consumers could peruse vases, as if in a museum, and see them set out as table settings in model dining rooms.

Wedgwood collaborated with Lunar Society friends to bring unique British minerals to bear on his pottery, explore the latest chemistry, and join in enlightened conversation. He consumed Priestley's notes on electricity, and together they experimented with electroplating ceramics. Wedgwood filled his notebooks with comments on Priestley's books on chemistry, experimenting on vacuums, combustion, colors, and heat.

During the period 1767–1774, Wedgwood received mineral samples from Priestley, Darwin, Boulton's toy business partner John Fothergill, Bentley, and canal builder Brindley. "Have I ever told you about the wonderfull & surprising curiositys we find in our Navigation!" Wedgwood wrote to

Figure 4
Manual operation of the valve gear on an 1820 Watt pumping engine.
Photo by the author, London Museum of Water & Steam.

Bentley about the canal excavations. "Last month was found
under a bed of Clay, at the depth of five yards from the sur-
face, a prodicious rib, with the vertebre of the back bone of
a monstrous sized Fish, thought by some connoisseurs to
belong to the identical Whale that was so long ago swallowed
by Jonah!"

Lunar Society member John Whitehurst came to Etruria
to help design and build the factory but soon began sending

Wedgwood earths and clays from canal digging. From these researches Wedgwood produced the famous Jasperware, the hard, delicate blue or green ceramic that remains the most recognizable Wedgwood pottery.

In its time, Wedgwood's pottery resembled the modern semiconductor industry: imprinting fine patterns on delicate materials, heating them at high temperatures with precision and uniformity, coating in layers, producing intricate objects capable of carrying cultural information. If Wedgwood had read *The Wealth of Nations*, he might have grown despondent about ever competing with China, due to the nation's ancient comparative advantage in pottery of the highest quality (Marco Polo actually coined the term *porcelain* in the thirteenth century, because the Chinese white ceramic reminded him of small white Italian pigs, *porcella*).

Instead, Wedgwood started with the humble, low-quality products in his hometown and built a local manufactory, a workforce, and a transportation network that grew to lead the global industry. Process innovation, in the new glazes and techniques, combined with product innovation, in new shape and designs, flexibly tuned to local taste and demand.

GENERATIONS OF INDUSTRIAL TRANSFORMATION

The last couple of decades have seen rapid changes in electronics and software. No sooner had the iPhone come out in 2006 than consumers were treated to rapid updates in succession, with ever increasing capability and connectivity (though the telephone network was already more than a hundred years old, and people had been experimenting with mobile computing devices for decades).

In industrial systems, however, things change more slowly. Even when a technology has been invented, developed, and proven, it can still take decades to transform a system. Often, these systems have safety or reliability dimensions that slow switchover. You don't want to try a new source of electricity only to find that it fails when the weather gets cold.

Thomas Edison introduced the first practical electric lighting system, the Pearl Street Station, in 1882. It was another fifty years before half of American households had electricity. And that was to build a system where no legacy technology, industry, or regulation had existed before (excepting perhaps gas lighting).

Consider aviation, the most dramatic technological change of the early twentieth century. From the day the Wright brothers first flew, it was thirty years before the Douglas Aircraft Company created an airliner, the DC-3, that could actually fly passengers economically and make a profit. It took another thirty years before jet travel as we know it today became feasible.

A rule of thumb, then, for transforming industrial systems is that it takes about thirty years. Not thirty years from invention to full deployment, but thirty years after a technology begins to proliferate for it to fully transform a system. This guide has the advantage of about matching a generation and an adult career cycle.

It also means that whatever technologies must decarbonize industry by 2050 better be largely here today. Current work in the lab getting published as science papers may generate promising results, but it needs to hurry if it's going transform large systems by 2050. By contrast, industry should support technologies that are beginning to find their way into deployment and adoption today.

I have been turning models, and preparing to make such *Machines of the Men* as cannot Err.
—Josiah Wedgwood to Thomas Bentley, 1769

Etruria opened its doors in 1769. To celebrate the first day of production, Wedgwood entered the pottery in front of a crowd of workers and onlookers. He sat down at a potter's wheel and personally threw six black basalt Etruscan-style pots while Bentley turned the wheel. Of the six, two cracked during firing; the four that emerged intact were sent to London for painting. Known as the "first day vases," they survive in collections today as among the finest examples of the Wedgwood black basalt.

This performance of individual craftsmanship was a source of Wedgwood's power. Indeed, his personal and business lives were inseparable, as were his dual identities as a craftsman and as a business leader. He built housing for hundreds of workers at the factory and built his own manor house there too.

Despite the domesticity, work at the factory hummed with industrial rhythms. Wedgwood had a vision of the entire factory as a machine, including people, equipment, and architecture. He organized the very factory site along these lines: separating the "useful works" and the "ornamental works" in different buildings—embedding into the landscape the division of labor, the years of experiments, and the secret recipes, in five separate structures.

Clay traveled in a circle: raw materials initially unloaded at the canal, and the finished products ending back there for shipping. In between they progressed in sequence from painting, firing in kilns, recording, to storage (sometimes to further fine painting in London).

The division of labor followed suit. Wedgwood thought that workers should master either coarse or fine pottery and not mix the two. "The 'fine figure' Painters are another order of beings," he wrote, "compared with the common 'flower painters.'" Each worker should specialize in their task, Wedgwood dictated, as "there is no such thing as making now & then a few of any article to have them tolerable [in quality]."

Wedgwood's secretary recorded an inventory of trades at the factory: a few famous modelers (who made the actual sculptural forms) like William Hackwood, and numerous painters, grinders, printers, liners, borders. For Jasperware, there were ornamenters, turners, slip makers, grinders, scourers and mold makers, turners, throwers, handlers, firemen, overlooks and packers, spout makers, engravers, polishers, sorters, dippers, brushers, and coopers. Of nearly three hundred workers at Etruria, only five lacked a job description. These five were listed as "odd men."

A film of Wedgwood production from 1950 (available on YouTube) shows the processes as virtually unchanged from these earliest days. Workers perform individual tasks on racks of uniform pieces, attach figures from standardized molds onto uniform pots, and hand carry the work from one station to another. Jigs and fixtures for handles and other items help support the division of labor and incrementally increase productivity. At least in the film, the workers are well dressed, the works clean and ordered. The only twentieth, or even nineteenth, century additions seem to be electric kilns and electric light.

Wedgwood found the older potters less malleable than younger ones, though more because of the types of products than the nature of the work. "Some will not work in Black. Others say they shall never learn this new business & want to be release'd to make Terrines & sau(ce) boats again." He preferred to train workers from youth to the new methods and the new industrial culture. In 1790, nearly 25 percent of his workers were apprentices, many of them young women.

A set of "Potters' Instructions" from 1780 describes every aspect of the new habits of mind; in modern terms these rules are the software of the Wedgwood operating system. The factory instituted a clocking-in system with paper tickets, where workers would log their entry and exit from the factory. Clerks and inspectors, who today would be called call managers, enforced the rules. A "price book of workmanship" sought to identify all sources of value—and waste—in the processes, an early cost accounting. "Rent goes on," he wrote to Bentley, "whether we do much or little in the time." Factory principles focused on the elimination of waste—of time, of clay, of labor.

Factory rules prohibited "any workman conveying Ale or Licquor into the manufactory," or "any person writing obscene or other writing upon the walls." Fornicating under the stairs: definitely out of bounds. The company issued fines for lateness and drinking.

Wedgwood personally enforced quality, with management methods we'd frown upon today. It was not unknown for him patrol the shops and smash imperfect pots with his cane. "My name has ben made such a scarecrow to them," he wrote of his workers, "that the poor fellows are frighten'd out of their wits when they hear of Mr. W. coming to town."

Wedgwood found himself constantly at odds with what he found to be poor work habits. Three-day drinking sprees or parish "wakes" (fairs) routinely interrupted work. He constantly complained of "dilatory, drunken, Idle, worthless workmen," of the "devout regard for Saint Monday" as workers drank through their prior week's wages. At the center of the Etruria campus was a bell to keep time.

What Wedgwood saw as weak work ethics the workers themselves likely saw as traditional ways of working, not yet tuned to the clock time and regularity of industrial production. They were experiencing a reshaping of human work to regular rhythms, at the start of what would take generations to transform. They were also experiencing what would be an enduring conflict: between the precision, regularity, and standardization favored by new technical systems and the ordinary variability of human beings and behavior. The problem changes through the generations, and with changes in technology, but dogs us still.

Wedgwood tried to operate in his employees' interests. He emphasized sanitation among the workers and shops. Lead poisoning was a known problem. Indeed, later accounts by Friedrich Engels would point to lead poisoning in the pottery industry as a source of oppression (though Engels did not mention Wedgwood). "The utmost cleanness should be observed thro'out the whole slip & clay house," read the rules at Etruria, "the floors kept clean—& (even) *the avenues leading to the slip & clay houses should be kept clean likewise.*" Similarly, "The dipping rooms to be cleaned out with a mop *never* brushed . . . no one to be allowed to eat in the dipping room." Wedgwood provided protective clothing for the workers. They were to wear smocks in the dipping room and remove them when exiting. Wedgwood and Bentley were horrified when they received a report detecting traces of lead in their creamware—though it turned out not enough to be dangerous, as it was present in low levels and didn't come off in the food.

The rules and practices created a new industry. "To pretend as some do that the division of labor destroyed skill," writes one historian, "is to deny the superiority of Wedgwood's products over his rivals, and to sentimentalize the crude Staffordshire salt glaze of his predecessors." Division of labor focused skill and increased it. Wedgwood's goal was "to make *Artists* . . . [of] . . . mere men" and "to make such *machines* of the *Men* as cannot err."

Some historians look back and see the work discipline as inherently oppressive, especially in light of the abuses that came afterward. But to do so simplistically risks romanticizing the life of the earlier craft-based work. In little more than a decade, Wedgwood had taken a downtrodden industry and community and made it the envy of the world.

Today we still value cleanliness and sobriety as foundations of human health. In the words of one historian, Wedgwood had "certainly produced a team of workmen who were cleaner, soberer, healthier, more careful, more punctual, more skilled and less wasteful than any other potter had produced before."

We don't have direct evidence of Wedgwood workers' experience, as they left few written records and didn't attract biographers. But the Wedgwood factory did not have strikes, and some workers surely welcomed the regular work and new living spaces. After fifteen years, Wedgwood had cultivated, trained, and created a disciplined workforce that competed with the best Chinese porcelain.

The division of labor found its way into Wedgwood's scientific experiments, one of which epitomizes his approach to industrial enlightenment. Around 1780, Wedgwood began experimenting with pyrometry, new methods to measure very high temperatures. A mercury thermometer can only measure up to the boiling point of mercury, 675 degrees Fahrenheit.

Wedgwood surveyed the published literature and began experimenting. He observed that above about 600 degrees, a piece of ceramic would shrink in proportion to the temperature. He then created a standardized set of ceramic cylinders about a half-inch in diameter and a brass or ceramic gauge to measure their shrinkage and size. The gauge had a converging diagonal shape, so the distance the cylinder would fit down the gauge was a measure of the temperature it had experienced in the kiln.

With his pyrometer, Wedgwood made his own scale of temperature and measured the boiling points of brass, copper, silver, and gold. He discussed these experiments with Boulton and Priestley, possibly at a Lunar Society meeting. After Wedgwood read a paper on the invention to the Royal Society in 1782, the illustrious society elected him a member.

Wedgwood made his own temperature scale, though he was unable to calibrate it accurately against the existing Fahrenheit scale. Nonetheless, improved methods did not arrive until decades later, so Wedgwood's scale was widely used. Wedgwood made dozens of his pyrometers for other researchers, including Joseph Black, Priestley, and Antoine Lavoisier in Paris. He also made one for King George III.

It is unclear whether Wedgwood's workers used the pyrometer for measuring temperatures in the kilns during production, but that was Wedgwood's goal. Hence it offers one window into how Wedgwood viewed technology in relation to his workers' skills. Contemporary practice, and the domain of the master potter, was to assess the temperature of the kiln by eye, looking at the color of the pottery—red versus white and so on. The pyrometer began to attach numbers to those judgments. Instead of a master potter's assessment of when the firing was done, in Wedgwood's mind the process would eventually be formally described by temperature and time: fire the pots at this temperature for so many minutes.

Would a pyrometer deskill a potter? In some sense yes, as the pyrometer replaced the potter's judgment with a physical artifact. This object embodied the knowledge gained by Wedgwood in his lifetime of experience and translated it into numerical measures.

Yet simply moving this one assessment from eye to thing would not itself deskill future potters or kiln operators. Rather, those workers could now experiment with repeatability, program more complex temperature profiles for firing, and greatly expand the range of temperatures with which they could work. Wedgwood's pyrometer liberated his artisans from the craft measurement of heat, enabling them to master more precise and complex processes.

Someday we may be able to say the same thing about generative AI.

Where industrial culture in England had to grow within a traditional system focused on colonialism and trade, in America industrial production could be more tightly linked to founding republican ideals. The Industrial Enlightenment shaped the early United States, through Thomas Jefferson, Benjamin Franklin, and American artisanal culture.

Franklin, a Lunar Society member and the original American rags-to-riches story, built a character around his industry. Franklin's invention was as much himself and his public image as any business or piece of equipment. So, too, with Paul Revere.

Today you can go to the Museum of Fine Arts (MFA) in Boston and see a painting of Paul Revere (1734–1818) by John Singleton Copley, who usually painted clergy and gentry. This is the only portrait he made of an artisan. Though most famous for his midnight ride in the American Revolution (immortalized by Henry Wadsworth Longfellow's poem a century later), in his time Revere was a well-known silversmith.

In Copley's portrait, Revere holds one of his teapots, unfinished, wearing workmen's clothing. He sits surrounded by the tools he will use to engrave and finish the piece. The museum also displays Revere's silver work: sugar bowls, pitchers, labels, and a spoon. On Revere's "Sons of Liberty Bowl," also in the

MFA collection, he inscribed the names of Massachusetts patriots who criticized the English Townshend Acts of 1767 that levied taxies on goods like glass, lead, and tea. For Revere, crafting silver could be a political act.

Revere's later silver work is designed with clean, modern lines in the Federal style—the American equivalent to British neoclassicism. The Revere silver comprises beautiful examples of American craft work from the period.

As one example, a tea service on display at the Minneapolis Institute of Art endures as one of the finest examples of Federal style silver from the eighteenth century United States. Revere made it in 1792 out of rolled silver sheet. The long, vertical flutes on the side of the teapot and sugar bowl allude to classical columns like those found on Boulton and Watt steam engines from the same time. The shape of the urn is inspired by Wedgwood-style classical pottery. The clean lines of the entire set reflect the cool early modernism of the American Enlightenment.

Born about the same time as Matthew Boulton, Revere began his career as an apprentice silversmith in Boston. He started his own shop in 1757, around the same time William Small moved to Birmingham. He was a likely reader of Franklin's *Poor Richard's Almanack* and absorbed its values of industry and frugality.

Revere's artisanal work brought him well into the heart of the Industrious Revolution, with early adoptions of wage labor, the division of labor, and subcontracting; his accounting practices newly emphasized profit and loss. His high-end work served the wealthy, but Revere also made a host of smaller goods, akin to Boulton's toys, for the middle classes.

Like Boulton, Watt, and Wedgwood, Revere achieved the status of the master craftsman, and like them he pushed further in a quest for professional development. Revere bridged the hands-on workshop with the consuming desires of the merchant and professional classes—part of what enabled him to play his famous role in the early revolution.

During the Revolutionary War, in addition to military activities, he helped set up a powder mill outside of Boston to supply gunpowder to the patriots (twenty-five years later, he would purchase that mill for his own industry). He introduced rolling machinery into his silver business and began to make novel products based on rolled, rather than hammered, silver plate. Revere embraced the division of labor and mechanization and the accompanying managerial skills it took to make them effective.

After the war, American manufacturing boomed to serve pent-up demand and a relative scarcity of British goods. Revere moved from fine silver to higher-volume toys and small metal objects. He opened a foundry in Boston's North End and advertised, "A general assortment of hardware, consisting of Brass, Copper, Pewter, Ironmongery, and Plated Wares." Like a New World Matthew Boulton, his firm sold tea and coffee urns, spoons, buckles, and trim, "the whole done equal to any imported, and upon the lowest terms." Though the new shop was a success, Revere pushed further to industrial scale.

Revere began casting bells in 1792, and his firm cast twenty-seven during the period of Revere's involvement (it would eventually cast more than nine hundred, of which almost 150 survive today). Engineer-historian Robert Martello has ably chronicled Revere's industrial transformation. "Beginning with that first bell," writes Martello, "he realized that the makers of quality items played their own vital role in the history of the new nation. . . . He came to realize that the foundry oven melded the characteristics of tools and machines: it required skilled labor and could be used in a flexible manner to produce different products, but an expert could produce consistent output by following a standard set of production practices."

To master bell making, Revere learned a host of new skills to plan production, develop products, and source and acquire raw material. He helped his workers learn to create and pour molds. With each skill he drew on his artisanal background but also ventured into new territory.

The techniques for casting bells resembled those for cannons. Each product served the community in a different way—bells to bring together a town for church or work, cannons to defend the nascent nation. "During his iron-casting career Revere taught himself and his employees to use the foundry oven in an even more machinelike way," writes Martello, "optimizing its use until they could produce highly standardized output." The federal government ordered ten cannons in 1794. Revere was elected the first president of the Boston Mechanic Organization (later called the Massachusetts Charitable Mechanics Association) in 1795, a group of like-minded artisans who aimed to improve conditions for their fellow workers in the new era.

In each of these markets Revere was competing with British goods, which at least at first were available at higher quality and comparable cost. Gradually, Revere's products improved to the point where he proudly compared them to the quality of imports. He also did not have the delays and hazards associated with transatlantic shipping and, like Wedgwood, was close to his customers.

Revere then moved into copper production. He started with the basic items required for shipbuilding: bolts, spikes, and the numerous other fasteners required to outfit and hold together a large wooden ship. Drawing on his silver-working skills, he developed techniques to heat and work the copper to make it strong and durable. Like Wedgwood, he was an experimenter. He drew on some basic scientific principles to guide his experiments—understanding, for example, the grain structure of metals and how the various operations shaped those grains. During the 1790s, he made the fittings for the USS *Constitution*, delivering a total of more than eight thousand pounds of copper products for the ship.

The next challenge in copper work was rolling sheets, which turned out to be a key strategic technology. The British navy was finding that copper sheets applied to a ship's hull could reduce biofouling—the accumulation of algae, barnacles, and other marine life that would grow under a ship's hull and harm its speed and maneuverability. Cladding a ship with copper proved to be one of the few effective deterrents. Copper products came to represent about 15 percent of a ship's total cost, more than half of which became copper sheeting.

Revere knew about the British copper cladding, and his timing was opportune, for the last years of the century corresponded with a desire to domestically build a navy to protect American interests at sea. The Department of the Navy was founded in 1798, and its first secretary, Benjamin Stoddert, sought ways to protect American vessels from French attacks. He established American navy yards and began building several frigates.

"You have it much at heart to finish all the Ships built for our government with copper from mines in the United States," Revere wrote to Stoddart in 1800, emphasizing domestic sourcing. But the government could not find someone to smelt the copper. "I have never tried," Revere admitted, but boldly offered, "from the experiments I have made I have no doubt I can do it." He needed the government to send him the raw material and pay him to build a furnace. Revere could also provide bolts and spikes for the new vessels.

Stoddert agreed, for he had tried several other US manufacturers without success. In 1800, the US government gave Revere a $10,000 loan to help establish a copper rolling mill. Now all Revere had to do was figure out how to smelt copper ore and roll it into sheets. He began by manually pounding the ore, smelting it in an inefficient furnace, and slowly rolling it with one of his old silver mills. He pulled it off: in 1801, he sent Stoddert a sample sheet of rolled copper. "It is one Evidence that Copper can be got in our own Country & manufactured into Materials for Ship Building."

The loan helped Revere purchase an iron mill in Canton, Massachusetts, on the site of the powder mill he had set up years earlier. The site came equipped with waterwheels, associated equipment, and even iron rollers that Revere began adapting to shape copper. The copper sheet began flowing from the rollers the following year. "We are daily gaining experience," he wrote in 1803 to ship designer Joshua Humphreys, "I cannot help acknowledging that I have done better than my expectations. Our sheets are as well finished and as soft & as free from scales & cannot be distinguished from English." He soon rolled enough copper for the navy to pay off the loan.

The success was short-lived, however, for when Thomas Jefferson was elected president in 1800, ambitious plans for the navy were shelved. Revere found himself at the epicenter of a nascent debate on American manufacturing—the same debate that shapes today's industrial strategy.

FROM REVERE TO R&D

As early as 1775, some urged the not-yet-born country to lessen dependence on foreign imports, an imperative strengthened by the Revolutionary War. The Springfield Armory was created soon after to manufacture rifles for the army. In 1790, President George Washington asserted that "a free people . . . should promote such manufactories as tend to render them independent of others for essential, particularly military, supplies." Congress then asked the secretary of the treasury to report on how that might be accomplished.

Though it didn't make it into his Broadway musical, in 1791 Secretary of the Treasury Alexander Hamilton wrote his famous "Report on Manufacturers" (simultaneous with his affair with Maria Reynolds). Its recommendations were not specifically enacted, but the document stands as an articulation of broad principles valuing industry and manufacturing in the early republic. It has underlain US policy toward industry for the past two hundred plus years and echoes through numerous successive policies. Tench Coxe was likely the coauthor.

Coxe, assistant secretary of the treasury under Hamilton, argued for the new federal government to support domestic manufacturing, "These wonderful machines, working as if they were animated beings, endowed with all the talents of their inventors, laboring with organs that never tire, and subject to no expense of food, or bed, or raiment, or dwelling, may be justly considered as equivalent to an immense body." Even these earliest framings of American policy embedded ideas of machines as tireless embodiments of their creators' skills, competing with the limitations of human workers. Machines in manufacturing could be operated by women and children, the line of thought went, without requiring male citizens to leave their farms.

Hamilton did not embrace Coxe's proposals, though he argued strenuously for the importance of industry and manufacturing in the early republic. Hamilton saw that human work to manipulate the physical world to create products that meet human needs and desires were core values that had shaped republican virtues since the country's inception.

Hamilton countered the argument that government should stay out of industry and let private industry develop on its own. He pointed to the extraordinary industrial moment that united the division of labor, "an extension of the use of Machinery," additional employment, immigration, and the need for roads, bridges, and canals. "Each of these circumstances," including military supplies, he concluded, "has a considerable influence upon the total mass of industrious effort in a community," adding "a degree of energy and effect." Hamilton helped found a Society for Establishing Useful Manufactures, which then with state support established Paterson, New Jersey, as an industrial city, employing water power for cotton mills.

George Washington signed the Naval Act of 1794, which authorized the construction of six frigates, including the USS *Constitution*, effectively creating the US Navy (the Continental Navy had relied on purchased vessels and had been disbanded after the Revolutionary War). That same year,

Congress established federal arsenals for military equipment, which Washington created at Harpers Ferry, Virginia, to complement the one in Springfield, Massachusetts.

Both of these arsenals would channel the world of the Lunar Society's mechanical members and pioneer interchangeable-parts production in the first half of the nineteenth century.

Jefferson, while a child of the Enlightenment, saw Hamilton's proposals as a further extension of state financial power and could not find support in the Constitution for government science (outside of the patent clause). He did, however, acknowledge the importance of industry for the country but preferred to support it through education and internal improvements (roads and canals), which he thought would not compete with private enterprise.

Here Jefferson built on a broad sense in the early republic that independence in manufacturing represented "a second war for independence from Britain." Then, as now, Americans broadly supported the importance of industry and manufacturing; the political differences arose about how best for the government to foster it.

Jefferson remained an Enlightenment techie. In 1786, he visited England and witnessed the Boulton and Watt engine at the Albion Mill, observing it as "simple, great, and likely to have extensive consequences" in America.

In an experiment to adapt manufacturing to the system of American slavery, in the 1790s Jefferson created a mechanized nail factory at Monticello. He staffed it with enslaved members of the Hemings family, including his own children, who worked under difficult conditions while he sometimes supervised and worked at the forge.

As president, Jefferson encouraged exploration under the commerce clause of the Constitution, sponsoring the Lewis and Clark Expedition, which explored the newly acquired western territory after the Louisiana purchase, out of army funds. His instructions for the expedition would have made his teacher William Small proud: astronomical observation, natural history, and a heavy emphasis on recordkeeping. He also established the United States Military Academy at West Point in 1802, America's first engineering school, on clear Enlightenment principles (though those inherited from the French artillery corps). West Pointers would survey the land for railroads and found the American engineering profession.

National defense has been America's longest and most effective industrial policy.

Revere's copper reclad the hull of the USS *Constitution* in 1803. By 1804, Revere had moved all his operations to Canton. He employed around ten full-time laborers in his new factory. Breaking with traditional practice, he did not bring them on as apprentices, but rather paid them as wage labor. That year, his son embarked on a tour of Europe to view factories and learn how they operated, including visiting the Birmingham toy industry (likely including Soho). He was impressed with what he learned, and the European factories influenced the layout of the Revere Canton works.

Jefferson's policies hurt Revere by embargoing a host of European imports in hopes of protecting American shipping while the British and French were at war. The embargoes were ineffective, but they did stifle American manufacturing by limiting trade, and severely hampered Revere's copper rolling. Revere sought to fill the gap by making copper boilers for stills, and his copper covered the dome of the Massachusetts state house.

By the time Revere retired in 1811, his site in Canton housed a water-powered rolling mill, two furnaces, boring machinery, and lathes for cannon. He had now formed a corporation and brought his son into the business. When the War of 1812 broke out, the country was prepared to provide copper sheeting to its new naval vessels, as the Revere works were able to supply several tons per week.

Revere also had inquiries that would point to another industrial future. "We are to supply some Gentlemen in New York," he recorded, "with upwards of 16,000 lb of [copper] sheet 3 feet wide by 5 feet long to make two boilers for two Steam Boats." Boilers needed more uniform copper than that used for sheathing and required thicker sheets to contain internal steam pressure.

The two "Gentlemen in New York" who came to Revere were Robert Livingston and Robert Fulton, who were building the first reliable steamship service between New York and Albany.

When their ship the *Clermont* first plied the Hudson River in 1807, driving its waterwheel was a modified Boulton and Watt steam engine, the first in the United States. Later versions would include Revere copper boilerplate, merging the Lunar Society–era steam technology with Revere's industrial republicanism. Revere would supply Livingston and Fulton with many tons of boilerplate in the coming years.

Yet this appearance of a Boulton and Watt engine on Fulton's steamboat was more a short-lived influence than the shape of things to come. Only a few Boulton and Watt engines would be erected in the United States, where abundant water power drove early industrialization.

In the United States, and increasingly in Britain, the frontier of steam was in mobility. Yet Boulton and Watt would not license their patent, which impeded improvement and experimentation for the last quarter of the century. Several high-profile litigations brought by the firm stifled innovation in the industry; Watt even discouraged Murdock from experimenting with locomotives in in his own shop.

In the United States, an engine beyond Watt's invention— the noncondensing, high-pressure engine developed by Oliver

Evans and John Stevens (and Richard Trevithick and others in England)—became dominant for steamboats. By the early nineteenth century, high-pressure engines were powering pumping stations and industrial sites as well. Compared to the lumbering, building-sized Boulton and Watt engines, high-pressure engines were smaller, faster, simpler, less expensive, and required less maintenance.

These benefits greatly improved the engines' flexibility, allowing users to creatively apply them in numerous settings. Trevithick and his partner John Bull used a high-pressure engine to make locomotives in the early 1800s, though the idea would not become practical for mobility until John and Robert Stephenson proved the *Rocket* at the Rainhill Trials in 1829. From there, industrial mobility would transform the century, in both countries.

James Watt not only missed this important development but actively resisted it. Even though he included high-pressure steam in the 1782 patent, he would not use, nor would he condone, steam of more than a few pounds of pressure (even proposing a resolution for Parliament to ban the practice). Reverence for his leadership was such that much of English practice followed suit. "Watt early reached a stage," writes historian Louis Hunter, "where his most distinctive characteristic was opposition to change." Numerous of Watt's most distinguished historians agree.

AN R&D SYSTEM ADRIFT

Stemming from Jefferson's time, the list of government's influences on American technology is too long to recount. The Erie Canal (sponsored by state, rather than federal, government) mirrored the Lunar Society's projects, and large-scale railroads, surveyed by US Army engineers, built on that legacy to connect a vast landscape. Interchangeable parts manufacturing, developed in federal armories over decades before the Civil War, proved key to the arms production that won the war and seeded booms in typewriters, sewing machines, and bicycles. Those techniques, and the companies and master mechanics who made them, spawned the US automobile industry.

A similar pattern repeated itself in the twentieth century when the US Army purchased the first aircraft from the Wright brothers, spurring the US aviation industry. These technologies and the production systems that made them helped the country mobilize for World War II. The postwar world saw a government–university–industry research and development system that created modern electronics, flew Americans to the moon, and sponsored technology development for the internet and today's computer and software industries.

But somewhere in recent decades, this system began to lose its way. It focused on product innovation to the exclusion of process innovation. It proved highly successful in creating advanced systems for the military, from GPS to smart bombs, some of which had commercial impact.

Military systems value performance over cost, and the American R&D system fell behind in process innovation for manufacturing. The 1980s saw the United States with the greatest military in the world, but losing to Japanese production in automobiles and consumer electronics. A vast R&D system, from universities to national laboratories, excelled at pushing the envelope in everything from software to quantum computing. But too often those technologies would be developed here and produced there. As mentioned earlier, the list is distressingly long: semiconductors, solar panels, batteries, electric vehicles.

The American system produced a great number of PhDs, but fewer and fewer of them were American citizens or immigrants who wanted to (or were able to) stay in America. The country underinvested in a workforce that could produce, deploy, and maintain the complex technologies it develops. US government agencies cannot hire enough technical talent to either adopt or thoughtfully regulate the technologies as they evolve.

American policy has focused on generating new technologies, and not nearly enough on the human contributions: adoption, safety, maintenance, infrastructure. The country has failed to incorporate workers into the story of technology development and grant them a share of improving productivity.

MANUFACTURING SOCIAL CHANGE

In the later decades of the eighteenth century, growing numbers of people began to question the unjust foundations of
England's budding industrial fortunes and the far-off consequences of making and buying things in a global system.

Quakers in particular criticized the gun trade, especially
as one of Birmingham's leading gun producers, Samuel Galton Jr., was a Quaker. A former student of Priestley's at the
Warrington dissenting academy, he worked in his own private
laboratory and became a Lunar Society member (likely the
wealthiest of the group). He defended his family's business
as supporting defense against war rather than war itself, and
professing no control over the final use of his products. The
Quakers rejected the argument and promptly expelled him
from their number.

An "anti-saccharite" movement arose to call attention to
the colonial and oppressive origins of the sugar so beloved by
the English. Sugar imports came from slave plantations. The
availability of capital for improved infrastructure like canals
was aided, though not necessarily enabled, by the thriving
international trades in both enslaved humans and the products of their labor.

Josiah Wedgwood lived within, and benefited from, an
English economy reliant on slavery. Yet he was at once an
anti-saccharite and a large-scale producer of sugar bowls.

Self-conscious about this uncomfortable contradiction, he joined the fight for reform. He supported both the American and French Revolutions. A Unitarian, Wedgwood joined the Society for the Abolition of the Slave Trade when it was founded, and soon joined its governing board. Boulton, Priestley, and Darwin all supported an early, though failed, parliamentary petition to raise anti-slavery funds.

Wedgwood contributed in the way he knew how: by making ceramics. In 1787, he created a medallion in jasper that depicted an enslaved man, in chains, on his knees, looking upward, with the words "Am I not a man and a brother?" It was likely modeled by William Hackwood, Wedgwood's chief modeler, after a design of Wedgwood's own.

A modern eye might see the image as submissive and stereotyped, even condescending. But the slave is speaking English, as much to the heavens as to the observer. The image is heroic, nonviolent.

Wedgwood's medallion became the emblem of the anti-slavery movement in Britain. The potter distributed thousands, likely donating rather than selling them, and thousands more were copied by others in various forms. Wedgwood sent boxes of them to abolitionist colleagues, including the movement's leader, Thomas Clarkson. Clarkson noted how quickly the medallions became a fashionable item with both women and men:

> Some had them inlaid in gold in the lids of their snuff-boxes. Of the ladies, some wore them in bracelets, and others had them fitted up in an ornamental manner as pins for their hair. At length the taste for wearing them became general, and thus a fashion, which usually confines itself to worthless things, was seen for once in the honourable office of promoting the cause of justice, humanity and freedom.

Clarkson credited the medallions with "turning the attention of our countrymen to the case of injured Africans and of procuring a warm interest in their favor." Erasmus Darwin used the image to illustrate his epic poem *Botanic Garden*. Wedgwood sent a package of his medallions to Benjamin Franklin; they became equally popular among early American abolitionists. The medallions even appeared in the American civil rights movement in the 1950s and 1960s.

Historians of abolition have described the medallion as "as a piece of propaganda central to the impassioned campaign for the ending of the transatlantic slave trade in the closing decades of the eighteenth century." Wedgwood biographer Tristram Hunt sees the medallion as "one of the most radical symbols in modern history," and one of Wedgwood's leading contributions.

Wedgwood's family continued to support the abolitionist cause after his death, and his son was elected to Parliament on an abolitionist platform. Britain abolished the slave trade in 1807, more than fifty years before the American Civil War.

Wedgwood synthesized process innovation (manufacturing) with product innovation (design). Through the Lunar Society, Wedgwood invested in multiple components of the pottery industry, from mining new minerals and understanding geology to funding canals for transport of his fragile wears. Being close to his market, Wedgewood understood the aristocracy's taste for revived classicism and middle-class women's desire to emulate those fashions. Wedgewood transformed industrial practices in his factories as extensions of scientific laboratory practices, though not without exposing his workers to toxic chemicals and despoiling the environment around his works. Despite, or because of, his success, he understood manufacturing could be an agent of social change.

"NO PHILOSOPHERS!"

Social change could cut both ways. On the afternoon of July 14, 1791, a mob gathered in Birmingham. They were incensed by a variety of events, including religious dissenters' campaign for equal rights and support for the French Revolution. The angry crowd gathered at the Royal Hotel, where guests were celebrating at a dinner to commemorate the storming of the Bastille two years before. The event was chaired by chemical industrialist and Lunar Society member James Keir.

Joseph Priestley found himself at the intersection of the mob's grievances; he had been warned and wisely stayed away. His scientific accomplishment matched only by his political radicalism, Priestley had recently not only supported the French Revolution but attacked the Anglican Church and Member of Parliament Edmund Burke for opposing it.

The mob harassed the dinner guests and ransacked the hotel. They then proceeded to attack meeting houses and homes of dissenters, beginning with Priestley's. The rioters angrily chased Priestley, who narrowly escaped. He watched from afar as the rioters trashed and then burned his home, destroying his library, his laboratory equipment, and years of scientific data. They tried, without success, to use one of his electrical machines to start the fire.

For days afterward, the roads in Birmingham were littered with Priestley's laboratory notes, floating in the breeze and trampled in the mud.

The riots continued for two days. The crowd attacked the house of Lunar Society member William Withering. Withering hired workers to defend the place and repelled the attack. Twenty-seven homes were looted or burned to the ground, including that of bookseller and historian William Hutton, who had first praised the Birmingham artisans as "full of industry." Boulton and Watt, fearing attacks on Soho, armed their workers, though their factory remained undisturbed. Finally soldiers arrived and dispersed the crowd.

The riot had multiple causes. It seems in part to have been preplanned, though the chaos became fueled by alchohol. Some of the agitators had been imported from London, supported by clergy and aristocrats. King George expressed satisfaction at Priestley's suffering. It became clear that the crowd singled out three groups of people for attack: Bastille dinner attendees, religious dissenters, and members of the Lunar Society.

Among the clashing mix of motives, the riot seems to have had some element of class conflict as artisans attacked the newly rich industrialists. Numerous rioters indeed were industrial artisans and laborers of Birmingham. The mob's chant of "No philosophers, Church and King forever!" did not go unnoticed by the Lunar Society members. Priestley barely escaped with his life.

Watt, generally growing more conservative as he approached wealth and retirement, lamented that after the riot, Birmingham grew "divided into two parties who hate one another mortally." He supported Priestley personally, though not his strident politics. When the Lunar Society next met, a couple of months after the riot, he grew concerned Priestley might attend and draw the attention of a new mob. As if to symbolize that the Lunar Society would never be the same, Watt brought a gun.

Priestley left the country and settled in Pennsylvania for the rest of his days. Though he corresponded with Thomas Jefferson on educational and political topics, Priestley never found intellectual companions nor scientific success as he had in Birmingham. Before he left England, he dedicated his last English scientific work to the Lunar Society. "There are few things I more regret," he wrote, "than the loss of your society. It both encouraged and enlightened me."

The Lunar Society continued to meet but would not achieve the vitality and fellowship of its previous years. The riots' disruption was not the only cause. Members were established, attending to their affairs as they aged, more recognized by the traditional London societies. As the century approached its end, an era closed as a new one opened.

The new era would bring promise and prosperity, threat and conflict. Boulton's and Watt's sons would develop the factory into a managerial enterprise. They would help usher in a world of even greater technical change and growth—alongside worker oppression—that the Lunar Society members could hardly have imagined. It would be decades before these changes improved the standard of living overall. Workers banded together to fight for reform, sometimes even smashing the machines, as industrial philosophies hardened into utilitarian ideologies. Romantic reaction included not only artists and poets but engineers themselves, seeking new ways achieve higher purpose through technical means.

Watt began withdrawing from the engine business around 1795, a wealthy man, and turned his attention to other tinkering and inventing. By this time the firm had sold nearly five hundred engines. While significant, this sum amounted to about one-quarter of the steam engines in Britain; half were still the Newcomen type, and others pirated from the patent.

Watt had struggled for decades to perfect his machine, and it was too much for him to see new ways to do things. He was tired, weakened by the deaths of his older children in 1794 and 1804 and the passing of other Lunar members: Darwin (1802), Priestley (1803), Robinson (1805), and Boulton (1809). Watt himself would pass in 1819.

Boulton and Watt's later conservatism does not detract from their accomplishments but merely underscores their humanity. Their vision was limited by their times, and the tools and workmen available to them. Industry is not a moment or an inventor but a layering of improvement over generations.

It is tempting to write profiles of young engineers and those working today to build new industrial systems to parallel the stories of Watt, Wedgwood, Boulton, and other Lunar Society members with compelling individual texture. Indeed, such new industrialists are working today, improving approaches to industrial problems that have been difficult to tackle in the past. I have interviewed and discussed the ideas in this book with dozens of them.

But Western culture has had two hundred and fifty years to curate stories of Lunar Society members. Scholars have pored over their writings, their machines, the details of their businesses. Knowing the outcomes gives the benefits of historical hindsight to understand the crucial moves, the essential collaborations, the people whose work went on to shape generations. All this by essentially reading their mail, their most private personal and business correspondence, and critically examining the machines they built.

One cannot write about the present, much less the future, in the same ways one does about the past. Were I to profile such people today, some of their businesses would surely fail by the time you read this book. Others who seem peripheral today will become wildly successful. We don't have access to the information within their businesses that will determine whether they succeed or fail, nor to their letters to spouses, children, and business partners.

Were I to gain access that data, I could not publish it. Were I to write about their businesses, I'd be reporting from the outside and end up with the public stories that entrepreneurs and journalists tell about their work rather than the gritty realities and glimpses of their inner lives.

Certainly, some technologies that will shape the world by 2050 are with us today, but which ones? Technologies like AI will shape industrial systems in profound ways in years to come, but few today can say what the killer apps will be. And certainly the socioeconomic and political conditions they are working within will change.

In fact, it is the future Boultons, Watts, and Maria Edgeworths who I hope are reading this book, who will tell us what the future will be as they make it themselves. For them, we should refashion the term *industrialist*: rather than connoting the extractive capitalists of the Andrew Carnegie era, it should describe skilled people applying their best years to industrial pursuits.

LUNAR SOCIETIES TODAY

Let's look at some places where new industrialisms are forming. For the past serval years I have been convening a group of industrial thinkers for series of dinners in a group I call the Lunar Society. The naming after the original is modest, aspirational toward a new Industrial Enlightenment. Like the original, today's Lunar Society has no formal status of any kind. Discussion topics have included new fields such as additive manufacturing, robotics, AI, and modeling our electrical grid. Attendees include founders of companies in manufacturing, computer-aided design, and renewable energy. The group also includes investors, policy makers, and frontline workers. I draw on these conversations below (though without direct attribution, based on the terms of discussion).

One of the new Lunar Society members founded and leads a novel power company that spun out of MIT. They have taken a rapid, pragmatic approach to building reactors as opposed to the larger government efforts that focus on more elegant, far-off solutions. It will take a few more years to prove the principle and will not be a panacea, but it could be an indispensable component of a clean energy future.

"People would be horrified if they really understood how things are still made today," one of our attendees said about manufacturing. Other countries with advanced manufacturing "make the US middle market look like the Stone Age," and have been more successful in upgrading their middle-market suppliers.

Too few students in universities and community colleges aspire to work in manufacturing. "When you speak to a classroom of students and ask, 'Who wants to work in manufacturing?' no one raises their hands. Compare this to AI." Yet an inability to manufacture new equipment will limit the ability to decarbonize our systems and to capture the benefits from making those systems.

Contrast manufacturing with software, where a rich open source movement provides everything from robotics algorithms to complete operating systems. Each user can grab what they want from the cloud, assemble it uniquely for their own purposes, and create a viable system.

In manufacturing, no such open source movement exists. Each factory has to reinvent process for itself, unable to draw on the experience of related firms, even those that are not competitors. Large manufacturers have little incentive to solve problems outside their existing operations or share their knowledge. Small and medium-sized manufacturers (fewer than 500 employees) in the United States make up about half of the ecosystem; they are undercapitalized, insufficiently digital, and unable to adopt new technologies.

Yet manufacturing knowledge in a small firm, even in a startup, resembles an operating system for a computer. The know-how forms connections between specialized elements that make the larger system achieve its purpose. To write software, anyone can go online and instantly access thousands of algorithms in Python or other languages. Yet there's no common repository for manufacturing knowledge. Where is the GitHub for manufacturing operations? Most robots today use different, custom interfaces that companies need to integrate from scratch. A little interoperability will go a long way.

Additive manufacturing (colloquially known as 3D printing) offers promise that the knowledge required to create a part is embedded in the digital instructions to drive the printer (similar to the idea behind numerically controlled machine tools two generations ago). This digital source resembles the recipes, factory layouts, and equipment that structured and enabled the skills of Boulton's or Wedgwood's workers. Design data encodes not on the shape but also instructions for manufacturing. Product innovation meets process innovation.

None of this is to say that all US manufacturers are narrow or lagging. Numerous companies do exceedingly well, compete in global markets, and create dignified sustainable work for their employees. But the system is undergoing generational change, in both people and technology, and may miss out on transformation.

Mobility faces a similar dilemma. You can't talk about the future of work without talking about how people get to work. Half of urban residents don't have convenient access to public transportation. Cities favor big capital projects over the thousands of small process innovations that improve the experience for riders.

Mobility is always a systems problem, combining different modes of transport, sometimes in one trip, sometimes on different days. Transit systems are the ultimate complex sociotechnical systems, focused on moving large numbers of people from one place (usually home) to another (usually work).

Others are trying to electrify air transport, a challenging, if essential, transition. Electric aircraft are just coming off the drawing boards and won't be able to directly compete with fossil-fueled aircraft for the foreseeable future (or possibly ever). Hybrid and other novel aircraft may have a better shot. For shorter trips, which make up a large component of traffic, they may be useful in fixing a broken aerial mobility system. Any such vehicles will need to be manufactured in new kinds of factories with new kinds of workers.

Industrial transformation does not always require new technology but can often make better use of the technologies already in use. The COVID-era supply chain crisis, for example, was partly solved by extending the hours for unloading at American ports, as well as making it easier for truckers to identify and find the loads they were supposed to retrieve (a solution that helped labor as much as the systems' flows).

Here the United States suffers from concentration of industry into an uncompetitive system. Rail, shipping, ports, and docks are characterized by duopolies, monopolies, and other entrenched structures. In these cases, improved organization of existing work can yield gains that exceed those from new technologies and open the systems for future improvement. Companies are learning to "stress-test" their supply chains and understand where they can invest to bolster resilience.

New industrialists are improving urban transit systems to make them safer and more robust. Others are settling in for a decade-long push to make nuclear power practical and safe—not knowing whether it can be done. Small companies are inventing ways to draw green electrons from the immense power of the oceans to help fuel the developing world. Others look to create clean, quiet airplanes that run on stored electricity—such a difficult problem today that it will require rethinking the overall aviation system and how it moves things and people from there to here. Some are creating software that enables small, family-owned machine shops to better connect with each other and their customers. Others are devising electric motors to run cars and other prosaic machinery with new efficiency. Still others are looking for ways to merge the digital and material worlds through new techniques of additive manufacturing or materials.

Often these engineers and founders have working-class roots, as children of factory workers who went to college and studied engineering. Rather than go into fully abstract professions like law or finance (or economics), they chose to learn how to build things and to bring the lessons of abstraction learned in colleges and on computers into the material world.

The sum total of this work, over time, amounts to a new industrialism if it elevates its cultural status into a movement that values the material basis of our lives and seeks to improve them, literally from the ground up. None of these will be apps that scale in a year or two and sell to growing digital monopolies. None will scale with mass updates upon millions of devices or in the cloud in a matter of months. All will take years, probably decades, to prove themselves, to demonstrate their safety, to find users and adopters, and to transform the systems that supply our lives.

THE NEW INDUSTRIALISM

What should a new industrialism look like? How can we update the Industrial Enlightenment to the demands of today?

A new industrialism depends on public recognition of the intimate dependence of our lives on industrial systems and on a public that takes ownership of these systems and their consequences, rather than shifting blame to others or abstractions. "Technology" doesn't do anything in our world. It is people working with technologies who make choices and have impacts. The Maintainers, self-proclaimed Makers, and advocates for Right to Repair and Repair Rather than Replace aim to shift our relationships with the material world. These social movements seek technological change that focuses on adoption and maintenance, designs for long-term life and replacement, bringing agency back to people and away from abstractions like "technology."

A new defense industrial strategy in echoes these concerns. The return of battlefield attrition, and even trench warfare, in Ukraine apparently surprised planners with the notion that future wars will require not only software and intelligent systems but production lines as well. The manufacturing dimension of defense shrank considerably after the end of the Cold War. Lockheed Martin, Northrop Grumman, and Boeing, for example, are three companies that used to be six (actually many more). It will take considerable time to build a modernized defense industry, one largely but not exclusively domestic.

The US Defense Production Act, an old but little-known (in peacetime) law enabling government direction of industrial capacity, was invoked in 2021 for electronics, batteries, and even COVID vaccines. The Department of Defense is beginning to support not only the performance side of military equipment but production technologies as well, including resilient supply chains. In a familiar list, the defense industrial strategy supports expanded domestic production, diversified supplier bases, new production methods, data analytics to identify supply chain risks, and investments in supply chain resilience. Key to this new industrialism is upgrading and expanding the defense industrial workforce, including a program to destigmatize industrial careers.

A new industrialism builds a national culture that values regular application of work to the vital tasks of improving the world. The Lunar Society shows us how technological change evolves not linearly from breakthroughs in laboratories but rather through numerous interactions between people with different kinds of knowledge and skills and the material world. These include research and development (though more development than research) but also collaboration, maintenance, manufacturing, and other industrial aptitudes. Familiarity with the material world, not to mention mechanical skills, drove the insights of Boulton, Watt, and Wedgwood.

A new industrialism requires economics that can account for the indelibly human dimensions of technology: learning, skill, communication, and collaboration. Economists are beginning to realize what historians have known for decades: technology is a human product, an expression of numerous layers of human experience and the natural world, the embodiment of human knowledge. The future is not something that happens to us but rather something that people shape. Technological change is rife with human decision points; determining which technologies thrive, and how they are applied, results from numerous choices along the way.

Economists' laudable, if late, realization comes at a crucial moment, because how AI affects workers is not some inevitable result of the technology but a product of how those technologies are developed and deployed. "Generative AI will surely impact inequality," writes David Autor and colleagues, "but the nature of that effect depends on exactly how this technology is developed and applied," calling for emphasizing applications that augment human workers.

A recent conference at MIT on AI, "Shaping the Future of Work," brought together economists who focus on this question, noting that "what the role of AI and the future of work is what we will make it be." While the AI scientists who attended expressed typical technological enthusiasms, the economists were keen to discuss the statistics and concepts of augmentation versus automation and anxiety about automation—age-old issues. But the group was still unclear on what actual "shaping" would mean.

The Lunar Society offers numerous examples of how people shape technology: Queen Charlotte's state patronage of Wedgwood's creamware, Watt's attempt to repair the university's steam engine model, Boulton's search for better power for his Soho factory, Wilkinson's boring of the engine cylinders, even Watt's opposition to high pressure engines. Each of these actions, and countless others, shaped the eventual engines and industry. Today's government policies, grant announcements, maker spaces, engineering CAD systems, 3D printers, AI databases, and the stories we tell about all of them all shape the industries of tomorrow. So do the adoption decisions of transit systems, the deployment plans of managers, and the acceptance (or rejection) of new technologies by frontline workers and consumers.

New industrialism requires venture capital tuned to the life cycles of industrial technologies. The venture capital industry in the United States has been remarkably successful in creating and supporting transformational companies and technology.

But it has been more effective in software systems, business services, healthcare, and consumer products than in transforming industrial systems. PitchBook, one of the leading sources of data on venture deals, doesn't even track industrials as a category. Founders of companies seeking to decarbonize industrial systems are often left to cobble together government support with a few far-sighted private investors. Deployment and adoption are slow where regulation, safety, and certification dominate.

Most venture firms operate with an investment vehicle called a fund, which has a seven- to ten-year life: a couple of years to raise the capital, a couple more to deploy it, and then a couple to harvest returns by selling companies or initial public offerings (IPOs). That leaves a few years for companies to mature. These cycles were based on ideas like technology S-curves that show early periods of rapid adoption and then a leveling out with maturity, as befitted software companies of the 1990s and early 2000s.

Yet this S-curve does not typically describe companies working on industrial transformation. It can take several years to go from idea to technology, and several more to prove that technology is viable within a system. Then the adoption cycle begins, and it can go on for a very long time. As I often say to industrial startups, "If you're right about this, you'll just be getting started in ten years." Rather than the leveling off with the typical software S-curve, industrial adoption can experience compounding growth for decades.

Some venture firms are beginning to speak about "industrial dynamism," "American dynamism," or "industrial evolution," and even "reindustrialization." In 2022, Trevor Zimmerman and I cofounded an investment firm with a financial model better suited to these industrial adoption cycles. We called it Unless, as a bit of an allusion to the last word of the Dr. Seuss story *The Lorax*, wherein the Lorax himself leaves a mark labeled with that word: "Unless someone like you cares a whole awful lot, nothing is going to get better."

Unless invests in industrial startups and aims to provide capital to them during the potentially long period of adoption. Its structure doesn't have a seven-year fund life, so it is not constrained to that cycle. The goal is to finance industrial companies, and then stay with them for the long haul as they achieve industrial scale. We don't do far-out "straight out of the lab" type technologies but those that are technically proven and ready for adoption over a long cycle. That said, they all require patience and perseverance as large customers adopt new technologies to full scale.

A new industrialism should be human-centric, recognizing the knowledge and value in work, and share the benefits with all who work. Of all the facets of new industrial thinking, the most common threads are calls to incorporate people from all walks of life into the processes of technological change. Simply adding new machines or new software or new AIs will not achieve transformation, absent larger organizational changes.

The United States may be entering a golden age for workers. Demographic shifts and low immigration (not to mention the loss of a million workers during the pandemic) exacerbate labor scarcity. This pressure not only raises wages (and potentially inflation) but may lead employers to improve working conditions and invest in technology—to relieve workers of more mundane tasks and to focus on higher-value work. Labor statistics are revealing a post-pandemic wage compression that may be reversing the long trend toward inequality. Union victories at Amazon, Starbucks, in the auto industry, and elsewhere (and rising public support for unions) represent increased power that accompanies that scarcity.

Despite these gains, generations of underinvestment in workforce education, lopsided focus on college achievement, and underfunded community colleges have hollowed out tiers of the workforce that can be most effective for industrial improvement. Universities have grown increasingly disconnected from industrial problems and working life, furthering the divide between people who manipulate symbols and people who make things.

Glimmers of change are appearing. In popular culture, the maker movement of the past twenty years has elevated hands-on skills with machinery and materials as part of education, from elementary school to engineering school, beginning to create a generation of workers familiar with the satisfaction of making things.

One signal of a cultural shift is the DIY renaissance visible on YouTube. A simple search unveils a vast array of hands-on work easily accessible to any teenager with a laptop. Here one can find tutorials on machine tools, homemade jet engines, meticulously documented classic car restorations and boat-building, vacuum tube guitar amplifier restoration and design, homemade numerically controlled machine tools. Some display the wonder of modern industrial tools and processes—watching a five-axis mill create a propeller from a hunk of metal is hypnotizing and beautiful. Videos (and ancient training films) range from meticulous instruction to the informational, expository, sensational, and even dangerous.

Despite uneven quality, this material makes visible varieties of industrial work that amateurs have always done but were hidden in basements or local hobby clubs. Some may be boutique, hobbyist versions of a disappearing manufacturing sector—expensive motorcycles made as one-offs, collectors' items rather than mass-market products. Still, they highlight the human work that goes into making artifacts and the inherent fascination and satisfaction, both personal and social, of shaping and assembling materials.

Labor unions are beginning to recognize that they, too, can shape how technologies are deployed and adopted, rather than simply organizing to keep them out. For example, the AFL-CIO, which represents a huge number of American workers, now sponsors an innovation summit in parallel with the CES trade show in Las Vegas, featuring workers discussing adoption and deployment of new technologies.

Some are now beginning to ask about manufacturing after "lean"—the world-changing paradigm from the twentieth century (adapted from the Toyota Production System) that relies on continuous improvement and worker empowerment. One approach, "augmented lean," seeks to deploy digital technologies to help frontline workers solve problems as "a human centric framework for managing industrial operations" and to "transform frontline work toward knowledge work, and foster a renaissance in manufacturing." These ideas are promoted by the founder of a Boston company called Tulip, which provides a set of simplified IT tools ("no code") to empower frontline workers.

Others are calling for manufacturing to return to cities, to enable low-income workers better access to its quality jobs, and to integrate production with urban landscapes. "Bringing manufacturing jobs back to the center of a city," advocates argue, "could mitigate the negative effects of industrial sprawl and increase opportunities for diversifying the labor market." Other benefits include shorter commutes for urban workers, reduced delivery distances, and connecting production to the energy of the city. It is possible that new technologies such as additive manufacturing will enable more distributed systems of manufacturing (although in semiconductors, the trend has been the opposite, toward ever more expensive, centralized facilities).

Small and medium-sized firms constitute about half of the manufacturing in the United States. Yet they have been slow to adopt digital technologies, much less robotics and advanced techniques. Can digitalization help create good jobs at good wages?

Human-centered approaches are consistent with emerging best practices in industrial robotics, where the best companies seek improved flexibility and quality rather than simply labor replacement. Researchers are beginning to call for positive-sum automation which would simultaneously improve jobs and enhance productivity.

Proponents of the "good jobs" movement argue that, independent of technology, changes in managerial practice can make life better for workers and also enhance productivity. While some of those changes are simple—more repeatable schedules, less complexity in product offerings and sales—they do require systemic changes to last. Moreover, they are not dependent on new technology or devices—problems lie in the traditional social software of management practices, norms, and expectations.

Some are calling for new job categories to address gaps between frontline workers and engineers. A technologist career track would bridge the divide between operators, trained on specific machines, and engineers who understand processes and systems. As mentioned above, we should also revive the term *industrialist* by updating it from its nineteenth-century plutocratic image to the modern, practical worker thriving on the border between the digital and material worlds.

The pandemic dramatically showed how digital technology affords new ways to work, and new ways to go to work, though those ways are unevenly distributed. An age of AI will surely usher in new forms of working, still emergent. We are beyond the rhythm and tyranny of the clock: Not everyone needs to go to work in the same place at the same time every day. Digitally enabled work schedules can coexist with family schedules. Work can allow for periods of intense concentration and collaboration, and also for contemplation and focused solitude.

A great deal is happening from the ground up in industrial settings. Startups are finding new paradigms for trade work, moving beyond a union-based system of apprenticeship and hourly work toward more inclusive arrangements that offer training and steady trade work with career paths.

A new industrialism sees technology as a mediator in human relationships with the natural world. A systemic look at carbon emissions has created a new emphasis on material inputs to industrial systems like steel, cement, plastics, and ammonia. New startups have arisen to reduce carbon inputs to these and other industrial processes. An era of AI, despite its software promise, will inevitably draw on massive inputs of material resources and electrical energy – potentially even negating improvements in those sectors.

Indeed, startups and accelerators have emerged to focus on *hard tech*. The term conflates several issues: technologies that are difficult to develop (which certainly includes software), technologies that have significant hardware components (that is, things), and a troubling gender dimension. I don't find hard tech a useful descriptor, but some are using it as a stand-in for industrial technology. MIT has created a venture capital firm that focuses on "tough tech," or "deep tech," which avoid the gender problem but remain vague. In my experience with technology, I have not come across one that is not tough.

We have the buzzword we need: *industry*.

These components of new industrialism are not entirely new but rather a broad movement to be congealed and elevated. Thousands of companies at work today are experiments in new industrial life. They hire diverse workforces and support flexible schedules. They make extensive use of digital tools like simulation and online collaboration. They create novel organizational structures and ways of working and are committed to the moral, economic, and social value of making things in today's economy. Through them present-day Lunar Circles emerge, driving ideas of the new Industrial Enlightenment into technological change, tantalizing glimpses of possible futures.

COLLABORATIVE HEROES

The stories of the Lunar Society offer some specific lessons: the importance of big ideas in the Enlightenment, such as William Small's influence over Thomas Jefferson. The importance of the middle layer of artisan/engineers who stood between the working and upper classes and could be credible with both. The value of engaging the material world—Josiah Wedgwood throwing his first day vases in front of all his employees. The role of consumer goods in driving early manufacturing. The dangers of Adam Smith's oversimplified assumptions about factories. The importance of collaborative versus combative heroism. The value of auxiliary inventions—the boring machine, the double action, the parallel motion, the flyball governor, the Wedgwood pyrometer—to make new ideas work. The complementary emergence of steam engines and machine tools. The value of the prepared mind—Matthew Boulton recognizing James Watt when he met him, because Boulton had already experimented with steam. The importance of industry to American character and culture. And the importance of challenging conventions, as the members of Lunar Society did so successfully through dissent and nonconformity.

But the most important lessons are subtler dimensions of the Industrial Enlightenment. The Lunar Society's optimism was itself part of the threat it posed. Watt, Boulton,

Wedgwood, and the other Lunar Society members all differed in their technical skills, their business acumen, their human and industrial fates. They did, however, share an optimism for building the world, a belief that by collaborating and exchanging ideas, and translating those ideas into material artifacts, they could contribute to human welfare.

They were not utopians. None ever laid out in writing a picture of a future world where industry made life efficient and clean. None cached their inventive ideas in future certainties, "in the future we will . . ." Rather, they built things, improved processes, transformed old ways of doing things. But building things in new ways could be radical acts.

Then, as now, disease, inequality, substance abuse, economic stagnation, and political instability threatened a dark future. Yet these friends explored the sciences and ideas of the Enlightenment and brought them to the industrial arts.

We need not see them as extraordinary. The Lunar Society met as one of numerous intellectual gatherings in Britain at the time (although the one most relevant for industry). Numerous other mechanics, millwrights, and instrument makers improved and invented without the small group of friends. None of the members became known simply for their membership, but rather for their work.

In fact, the picture of industrial transformation deepens if we see the Lunar Society members as typical: talented (and limited) people whose industry and circumstances enabled extraordinary things. Among thousands of industrious people in a time of threatening political and social change, they chose to leave their marks by improving and building the world.

Were they heroes? Not in the Homeric sense, which survives today in Hollywood and tech media—conquering aristocrats creating eternal narratives through strength and victory. They were typical heroes—people who learn skills, work their ways up in life, latch on to new ideas, and strive to bring those ideas to fruition. They were collaborative heroes, who together explored the hidden behaviors of the material world and learned to use those behaviors to advantage.

My aim here is not to predict nor even to describe industrial futures, of which many are possible. Rather, the goal is to bring the thinking and speaking about those futures down to earth, figuratively and literally, to reinvigorate some of the ideals of the Industrial Enlightenment and, we hope, to avoid some of its pitfalls.

So for those budding Lunar Society members in today's small industrial companies, today's new industrialists, for those wishing to join them or create their own, or even for us ordinary consumers who shape technological change every time we ride in a car or use a computer, I suggest the following guidelines for industrial futures:

Marry product improvement to process improvement. Avoid utopian innovation speak. Think in systems. Emphasize adoption. Design for resilience and flexibility. Get excited about maintenance and repair. Value knowledge at every level of work. See the human intelligence embodied in every product and system. Form your own Lunar Societies to collaborate and explore.

As I was putting the finishing touches on this book, a total lunar eclipse captured the attention of America. I watched with old friends, on a farm in rural Vermont. The roads of New England were clogged with traffic as millions traveled to the path of totality to catch a glimpse of the shadow. All were struck by the sense of unity and wonder that accompanied the collective view, standing on a celestial body gazing out at our nearest neighbor.

That eclipse, and this book, suggests that the very phrase *lunar society* aptly describes a world we are building toward. A lunar society looks outward, from the legacy of the Apollo project to future explorations. The gaze sees the solar aura around the moon, or reflections of thousands of years of human mythology and aspiration, full of mystery but pock-marked by heavy bombardment. A face of light against a dark background, embodying natural cycles.

Hopefully, as it did for Maria Edgeworth, Matthew Boulton, Josiah Wedgwood, James and Ann Watt and their friends, that lunar reflection will light our way home.

ACKNOWLEDGMENTS

Writing a book like this builds on numerous conversations with colleagues over many years. At MIT, these have included David Autor, Suzanne Berger, Bill Bonvillian, John Hart, Simon Johnson, Chris Love, Victor McElheny, John Ochsendorf, Elisabeth Reynolds, Daniela Rus, Julie Shah, Merritt Roe Smith, Rosalind Williams, Dana Yoerger, and Jinhua Zhou. Generative AI was not used in the research or writing of the text.

I'm particularly grateful to the numerous founders and CEOs of new companies who are building industrial futures. These include, but are not limited to, Bret Boyd of Sustainment, Kyle Clarke of Beta Technologies, Ali Forsyth of Alloy, James Kinsey of Humatics, Natan Linder of Tulip, Bob Mumgaard of Commonwealth Fusion Systems, and Garth Sheldon-Coulson of Panthalassa. Other attendees at the new Lunar Society events have also provided wonderful perspectives, including Darren Aronofsky, Kate Darling, Julie Diop, Trey Foskett, Neil Lawrence, John Leonard, John McElheny, and Sree Ramaswamy. Trevor Zimmerman, my cofounder and partner at Unless, has proven an invaluable and patient partner in exploring these ideas and making them effective in the world. In that good work we are just getting started.

NOTES

CHAPTER 1

Page 1 Watt's fellow Scot, steel magnate Andrew Carnegie. Carnegie, *James Watt*.

Page 3 Watt's own memory half a century later. Dickinson and Jenkins, *James Watt*, cite the retelling in 1814, 23.

CHAPTER 2

Page 5 Ideals inherited from the Industrial Enlightenment. Mokyr, *The Enlightened Economy*; Jones, *Industrial Enlightenment*.

CHAPTER 3

Page 7 The big four. Smil, *How the World Really Works*.

Page 7 Two percent of US energy is consumed in the extraction of oil and gas. Griffith, *Electrify*, 41.

Page 8 Our food is partly made not just of oil. Smil, *How the World Really Works*, 217, 75.

CHAPTER 5

Page 11 Its buzzword "Industry 4.0." Schwab, *The Fourth Industrial Revolution*.

CHAPTER 6

Page 15 What historians have dubbed the "Industrial Enlightenment." Mokyr coined the term "Industrial Enlightenment," *The*

Enlightened Economy, 40; Jones, *Industrial Enlightenment*; Mokyr, *The Gifts of Athena*; Allen, *The British Industrial Revolution in Global Perspective*; Mokyr, *A Culture of Growth*.

Page 16 The Enlightenment has always had its critics. Horkheimer and Adorno, *Dialectic of Enlightenment*; Latour, *We Have Never Been Modern*.

Page 16 Historian Roy Porter argues. Porter, *Enlightenment*, xxii; Robertson, *The Enlightenment*.

Page 17 What historian Jenny Uglow said about the Lunar Society. Uglow, *The Lunar Men*, xiii.

CHAPTER 7

Page 19 A return to a more traditional way of doing things. Keilman, "America Is Back in the Factory Business."

Page 20 What some have called "innovation speak." Vinsel and Russel, *The Innovation Delusion*.

CHAPTER 8

Page 24 Birmingham tripled in size. Jones, *Industrial Enlightenment*, chapter 2.

Page 24 Guns (often destined for the slave trade). On the Birmingham gun trade, see Satia, *Empire of Guns*.

Page 25 This offspring of fancy. Quoted in Jones, *Industrial Enlightenment*, 42.

CHAPTER 9

Page 28 Mentions manufacturing only once. This analysis draws on Adler and Bonvillian, "America's Advanced Manufacturing Problem."

Page 28 To further scale, they move their operations overseas. Reynolds, Samuel, and Lawrence, "Learning by Building."

Page 29 Began fabricating chips for so many companies through its foundry model. Miller, *Chip War*.

CHAPTER 10

Page 32 Potentially one hundred times more productive than Smith reported. This analysis is based on Peaucelle and Guthrie, "How Adam Smith Found Inspiration in French Texts on Pin Making in the Eighteenth Century," with its detailed study of the sources of the *Encyclopedia* article, its errors, and Smith's misreadings.

Page 33 Gave economists a chance to show off their mathematical prowess. Krugman Quoted in Bonvillian, *Advanced Manufacturing*, 89.

Page 33 Whether mainstream economics can account for the critical role manufacturing plays. Bonvillian, *Advanced Manufacturing*, 69. For a detailed review of the failures of mainstream economics, see chapter 4.

Page 33 Limited range and subject matter. Deaton, *Economics in America*, 16, 233.

Page 34 Essentially alien to macroeconomics. Adler and Bonvillian, "America's Advanced Manufacturing Problem."

Page 34 Enabled the decline of manufacturing. Bonvillian, *Advanced Manufacturing*, 54–55; Berger, *Making in America*, 28–33.

Page 34 Concentrated in software, healthcare. Kauffman foundation report, quoted in Bonvillian, *Advanced Manufacturing*, 189.

Page 34 Key to a modern "service economy." Bonvillian, *Advanced Manufacturing*, 198; Smil, *Made in the USA*.

Page 34 Established structures, vested interests, high regulation, and high barriers to entry. Bonvillian, *Technological Innovation in Legacy Sectors*.

Page 35 Every manufacturing dollar and job engenders additional dollars and jobs. Bonvillian, *Advanced Manufacturing*, 61; Smil, *Made in the USA* offers a concise overview of the rise and decline of US manufacturing.

Page 35 Simultaneous management of tempo, production volume, and cost. Bonvillian, *Advanced Manufacturing*, 49.

CHAPTER 11

Page 38 Depend on specialized steels imported from China. Postel-wait, "Transformative Times"; US Department of Commerce, "The Effect of Imports of Transformers and Transformer Components on the National Security."

CHAPTER 12

Page 40 Was a master of every metallic art. Keir to Matthew Robinson Boulton, December 3, 1809. SoHo Archives, Birmingham Archives & Heritage at Birmingham Reference Library.

Page 41 We know it's matter. Boulton quoted in Uglow, "Matthew Boulton and the Lunar Society," in Mason, ed., *Matthew Boulton*, 7.

Page 41 Human experience and senses as key instruments for scientific observation. Riskin, *Science in the Age of Sensibility*.

Page 42 For we can both hear it and see it smell it & feel it. Boulton quoted in Uglow, "Matthew Boulton and the Lunar Society," in Mason, ed., *Matthew Boulton*, 7.

Page 42 Aligned with the most powerful forces of nature. Heilbron, *Electricity in the 17th and 18th Centuries*, chapter 14; Cohen, *Benjamin Franklin's Science*, chapter 6.

Page 43 He found electricity a curiosity. Van Doren quoted in Isaacson, *Benjamin Franklin*, 144.

Page 43 To a more provincial one based in philosophical societies. Musson and Robinson, *Science and Technology in the Industrial Revolution*, 89.

Page 45 A large waterwheel in one of the courts. Darwin, quoted in Demidowicz, *The Soho Manufactory, Mint and Foundry, West Midlands*, 44.

Page 45 Which like Bee hives crowded with the Sons of Industry. Quoted in Mason, ed., *Matthew Boulton*, 71.

Page 45 We passed an evening listening to him. Fanny De Luc quoted in Peter M. Jones, "Visitors to the Soho Factory," in Mason, ed., *Matthew Boulton*, 79.

Page 46 A prelude to the more organized Lunar Society. Schofield, *The Lunar Society of Birmingham*, 5.

Page 46 Could never legitimately claim the status of gentlemen Wood, *The Americanization of Benjamin Franklin*, 41.

Page 46 An article from 1803 that mentioned "English Worthies." Schofield, *The Lunar Society of Birmingham*, 5.

CHAPTER 13

Page 47 Draws on dozens of sites from Tier 1 suppliers. Elm Analytics, "Ever Wonder How Many US Auto Suppliers There Really Are?"; Sheffi, *The Magic Conveyor Belt*.

Page 48 EVs represent a new class of cyberphysical systems. Charette, "The EV Transition Explained."

Page 49 McKinsey consultants have labeled the US industrial ecosystem. Padhi, *The Titanium Economy*, 113, 207.

CHAPTER 14

Page 51 This use exploded. Williams, *Keywords*, 165.

Page 51 Worked with new energy and purpose. De Vries, *The Industrious Revolution*, 10, 32.

Page 51 An evolution from the natural cycles of the seasons. Mokyr, *The Enlightened Economy*, 272–275; Thompson, "Time, Work-Discipline, and Industrial Capitalism."

Page 52 Disproportionate numbers of industrial and commercial improvements. Mokyr, *The Enlightened Economy*, 362.

Page 52 Resembled the experiments made by the air pump. Priestley, *Lectures on History and General Policy*, 5, quoted in Schofield, *The Enlightenment of Joseph Priestley*, 125. On Priestley's readers and influence, see *The Enlightenment of Joseph Priestley*, 144.

Page 53 Pools of stagnant water. Priestley, *Letters to Pitt, 1787*, quoted in Musson and Robinson, *Science and Technology in the Industrial Revolution*, 167. On Warrington and Priestley's work there, see Schofield, *The Enlightenment of Joseph Priestley*, chapters 4, 5, 6.

Page 53 Cited by sociologist Max Weber. Weber, *The Protestant Ethic and the "Spirit" of Capitalism*, 48.

Page 53 By absorbing the gentility of the aristocracy and the work of the working class. Wood, *The Radicalism of the American Revolution*.

Page 54 A snobbery that would come to be shared by very disparate groups. Isaacson, *Benjamin Franklin*, 480.

CHAPTER 15

Page 55 There was plenty of toilet paper. If you think about it, the same number of rear ends need to be wiped, regardless of whether people are going to work or staying at home. Though there was a catch: toilet paper made for commercial applications is of a lighter grade than that marketed to consumers, so the commercial supply could not simply be adapted to domestic consumption.

Page 55 Producers stressed their systems to provide more without adding capacity. Helper and Soltas, "Why the Pandemic Has Disrupted Supply Chains."

Page 56 Ford had 40,000 F-150 trucks it could not ship. Sheffi, *The Magic Conveyor Belt*, 62.

Page 56 One of the first initiatives of the Biden administration. *Building Resilient Supply Chains, Revitalizing American Manufacturing, and Fostering Broad-Based Growth*.

CHAPTER 16

Page 57 The question is not why the item does not make it on time. Sheffi, *The Magic Conveyor Belt*, 19–26.

Page 58 Department stores created dream worlds. Williams, *Dream Worlds*.

Page 59 We have seen the supply chains and they are us. After defeating the British in the Battle of Lake Erie in the War of 1812, US Navy Master Commandant Oliver Perry wrote, "We have met the enemy and they are ours." In 1970, cartoonist Walt Kelley modified the quote for his character Pogo, for a poster for the first Earth Day poster. Pogo, confronted with a landscape of trash, says, "We have met the enemy and he is us." "'We Have Met the Enemy and He Is Us.'"

CHAPTER 17

Page 61 A biography of him appeared only in 2021. Clagett, *A Spark of Revolution*.

Page 61 They emphasized numbers and counting. Berry, *The Idea of Commercial Society in the Scottish Enlightenment*.

Page 62 For a couple of years Small served as the only professor. Clagett, *A Spark of Revolution*, 155–156.

Page 63 It was my great good fortune. Jefferson, *Autobiography*.

Page 63 Ostensibly he went to buy scientific instruments for the college. Clagett, *A Spark of Revolution*, 165, lists the instruments Small purchased.

Page 63 Franklin sent him to Birmingham. Schofield, *The Lunar Society*, 36–38.

CHAPTER 18

Page 65 *Conquer France by pottery ware*. Wedgwood to Bentley quoted in Smiles, *Josiah Wedgwood, His Personal History*, 148. Emphasis added.

Page 65 He actually coined the term *self help*. Smiles, *Self Help*; Hughes, "In His Proper Place."

Page 66 In part to reclaim his Americanness after living in Europe for decades. Wood, *The Americanization of Benjamin Franklin*.

Page 67 Joseph Campbell characterized the Western hero cycle. Campbell, *The Hero with a Thousand Faces*.

Page 67 A feminist critique of Campbell. Tatar, *The Heroine with 1001 Faces*.

CHAPTER 19

Page 69 The other Lunar Society manufacturer. There is some disagreement among historians about whether Wedgwood should be considered a member of the Lunar Society. But for our purposes here, given his close relationships with Boulton, Watt, Darwin, and others, his attendance at meetings, and his lengthy correspondence in the Lunar Circle, we will consider him one, as does Schofield, the preeminent Lunar Society historian.

Page 70 The enormous expense of the horse-power. Boulton quoted in Smiles, *Lives of Boulton and Watt*, 110.

Page 70 Newcomen engines made few demands on the manufacturing technology. Lindqvist, *Technology on Trial*, 110–116.

Page 71 He sent it to Benjamin Franklin for his opinion. Quoted in Schofield, *The Lunar Society of Birmingham*, 61.

CHAPTER 20

Page 73 MIT has commissioned a series of studies. Suzanne Berger, "MIT Research on Production," presentation at MIT Manufacturing Symposium, May 6, 2022.

Page 74 Jeff Wilke, an early graduate of Leaders for Manufacturing. Stone, *The Everything Store*, Kindle edition, loc. 2376, 2378, 2508.

Page 74 Later MIT studies included *Making in America*. Berger, *Making in America*; Locke, *Production in the Innovation Economy*.

Page 74 Including the Manufacturing USA institutes. MIT has also sponsored policy studies on the future of nuclear power, convergence in the life sciences, human spaceflight (which I led), and digital education. Thanks to Bill Bonvillian, retired head of MIT's Washington office, for filling in this history.

Page 74 Conversations included various jobs' "susceptibility to automation." Brynjolfsson and McAfee, *The Second Machine Age*; Frey and Osbourne, "The Future of Employment."

Page 75 Human roles in autonomous systems, in undersea exploration, aviation, and spaceflight. Mindell, *Our Robots, Ourselves*; Mindell, *Digital Apollo*.

Page 76 The task force published studies. Autor, Mindell, and Reynolds, "Research Report: MIT Work of the Future."

Page 77 Amazing digital tools that have transformed our desktops have offered the least to food service workers. Autor, Mindell, and Reynolds, *The Work of the Future*, 24, 45; Brooks, "Steps Toward Super Intelligence."

Page 77 More than 60 percent of the 2018 jobs. Autor et al., "New Frontiers."

Page 78 What Autor calls "wealth work." Autor, Mindell, and Reynolds, *The Work of the Future*, 14; Autor et al., "New Frontiers."

Page 78 The central challenge ahead. Autor, Mindell, and Reynolds, "The Work of the Future," 2.

Page 78 The "Work of the Future" study also concluded. Autor, Mindell, and Reynolds, *The Work of the Future*.

CHAPTER 21

Page 81 One can imagine the conversation as Watt filed at his vise, For a description of Watt's workshop, see Russell, *James Watt*, chapter 2. For the accounts of Watt and Boulton, and quotes from participants in what follows, I draw on Dickinson and Jenkins, *James Watt and the Steam Engine*, which was published on the Watt centenary and is a remarkably complete account of the partners, their process, and the details of the machines, including publication of a number of primary documents as appendices. Robinson and Musson's *James Watt and the Steam Revolution* offers numerous primary documents, and Dickinson's *James Watt* investigates the skills that enabled Watt's inventions. Smiles's *Lives of Boulton and Watt* laid out the basic narrative structure that most accounts follow. Russell's *James Watt* offers a more recent take, with emphasis on engineering cultures and a contemporary reading of artifacts and Watt's workshop. Other modern summary accounts include Marsden, *Watt's Perfect Engine*, and Rosen, *The Most Powerful Idea in the World*.

Page 82 The Watts had all struggled economically. For an economic history of the Watt family, see Jacob, *Scientific Culture and the Making of the Industrial West*, chapter 6.

Page 82 Watt and his family's connections to the slave trade. Davidson, "James Watt, Slavery, and Statues"; Mullen, *The Glasgow Sugar Aristocracy*, 78.

Page 82 Struggling as it were for life all through his childhood. Smiles, *Lives of Boulton and Watt*, 88.

Page 83 Served as a bridge between the university and local industry. Musson and Robinson, *Science and Technology in the Industrial Revolution*, 179.

Page 84 Black worked both as a professor and as an industrial consultant. Russell, *James Watt*; Musson and Robinson, *Science and Technology in the Industrial Revolution*.

Page 84 To follow the *clock metaphor* for nature. Shapin, *The Scientific Revolution*, 32.

Page 84 Clocks, machines, and their cousins, automata, seem self-acting and intelligent. On Enlightenment automata see Voskuhl, *Androids in the Enlightenment*, and Riskin, *The Restless Clock*.

Page 84 The Scientific Revolution's most important fact making machine. Shapin, *The Scientific Revolution*, 96.

Page 85 One Sunday in 1765 Watt went for a walk on the town green in Glasgow. This story was related by R. and J. Hart to the Glasgow Archaeological society in 1845, about a conversation they said they had with Watt in 1813 or 1814. Quoted in Dickinson and Jenkins, *James Watt and the Steam Engine*, 23. Hart's full account is reprinted in Robinson and Musson, eds., *James Watt and the Steam Revolution*, 40–45.

Page 87 We can see emerging, for the first time Pugh, in Baynes, *The Art of the Engineer*, 63. Drawings for Wilkinson's mill engines begin to specify 3D shape, including dimensions, but do not include tolerances. Watt used color coding and dimensions, according to Pugh in Baynes, *The Art of the Engineer*, 68–69; Boulton and Watt, "Directions for Erecting and Working the Newly-Invented Steam Engines," in Dickinson and Jensen, eds., *James Watt and the Steam Engine*. Beautiful reproductions of Watt's drawings can be found in the appendix of Robinson and Musson, *James Watt*.

Page 87 Sometimes the same drawing would stand for different sizes of engines. Dickinson and Jenkins, *James Watt and the Steam Engine*, 269.

CHAPTER 22

Page 91 A host of developments from World War II in electronics, computing, and automated weapons. Mindell, *Between Human and Machine*.

Page 92 The Apollo lunar landings. Mindell, *Digital Apollo*.

CHAPTER 23

Page 95 The greatest part of it must devolve on me who am from my natural inactivity. Watt to Boulton, October 11, 1768.

Page 96 To teach any blockhead in the nation to construct masterly engines. Small to Watt, February 5, 1769, reprinted in Robinson and Musson, *James Watt and the Steam Revolution*, 54–55.

Page 97 My idea was to settle a manufactory. Boulton to Watt, February 7, 1769, reprinted in Robinson and Musson, *James Watt and the Steam Revolution*, 62–63.

Page 99 He [Boulton] pleaded, he flattered, he pressured. Schofield, *The Lunar Society of Birmingham*, 332.

Page 99 what all the world desires to have. James Boswell, *Life of Dr. Johnson* (1791; Oxford 1998) quoted in McClean, "Introduction," in Mason, ed., *Matthew Boulton*, 6.

Page 99 So singular and so powerful a Machine. *Birmingham Gazette*, March 11, 1776, quoted in Dickinson and Jenkins, *James Watt and the Steam Engine*, 113.

CHAPTER 25

Page 106 Josiah Wedgwood became more integrated into the Lunar Circle. Schofield, *The Lunar Society of Birmingham*, 130.

Page 106 The first meeting called the "Lunar Society." Jones, *Industrial Enlightenment*, 89–94, has the most detailed accounting of the actual meetings.

Page 106 Bandy'd like a shuttlecock. Darwin to Boulton, April 5, 1778, quoted in Schofield, *The Lunar Society of Birmingham*, 143–144.

Page 107 Surely the best example of the eighteenth century, dissenting, lower middle-class scientist. Schofield, *The Lunar Society of Birmingham*, 198-9; Schofield, *The Enlightened Joseph Priestley*.

Page 108 Priestley later isolated and identified. Schofield, *The Lunar Society of Birmingham*, 197, has a brief summary of Priestley's chemical contributions.

Page 109 Watt deduced the composition of water. Schofield, *The Lunar Society of Birmingham*, 199; Schofield, *The Enlightened Joseph Priestley*, 172–4; Dickinson and Jenkins, *James Watt and the Steam Engine*, 59.

Page 109 The excitement of science and manufacturing. Uglow, *The Lunar Men*, xix.

Page 110 Also involved with his business ventures. Croft, "Of Material Service to Him."

Page 110 Maria corresponded with Lunar Society members. Chandler, "Edgeworth and Lunar Enlightenment"; also see Jacob, *Scientific Culture and the Making of the Industrial West*, 112–113.

Page 111 Wedgwood's canal building unearthed volumes of interesting fossils. Schofield, *The Lunar Society of Birmingham*, 162; Musson and Robinson, *Science and Technology in the Industrial Revolution*, 143.

Page 111 Priestley had cited Darwin's work. Schofield, *The Enlightened Joseph Priestley*, 150, 161.

Page 111 Lunar Society members kept commonplace books. Schofield, *The Lunar Society of Birmingham*, 167–168.

CHAPTER 26

Page 115 Much of the construction fell to millwrights. Musson and Robinson, *Science and Technology in the Industrial Revolution*,

98–99. Millwright quote from Fairbarn, *A Treatise on Mills and Millwork* (1861), quoted in Musson and Robinson, 429.

Page 116 Watt invented a counter. Schofield, *The Lunar Society of Birmingham*, 148.

Page 116 This strategy drove Watt to first quantify the horsepower. Dickinson and Jenkins, *James Watt and the Steam Engine*, 353–357, details Watt's quantifying the horsepower.

Page 117 There is no other Cornwall to be found. Quoted in Dickinson and Jenkins, *James Watt and the Steam Engine*, 56.

Page 117 In the evenings, he would arrange the fossils he collected. Smiles, *Lives of Boulton and Watt*, 284–285.

Page 118 It is all that I can do to keep him from sinking. Ann Watt to Matthew Boulton, quoted in Smiles, *Lives of Boulton and Watt*, 240–41.

Page 119 Bouton and Watt could pay off debts. Roll, *An Early Experiment in Industrial Organization*, 84, has figures based on commissions paid to Boulton and Watt's agent in Cornwall. Also see Dickinson and Jenkins, *James Watt and the Steam Engine*, 63.

Page 119 Steam mill mad. Boulton to Watt, June 21, 1781, reprinted in Robinson and Musson, *James Watt and the Steam Revolution*, 88.

Page 119 Of producing rotative motion. Smiles, *Lives of Boulton and Watt*, 260.

Page 119 Steam-powered mills would come to run two clocks side by side. Russell, *James Watt*, loc. 2602.

Page 120 Watt adapted a mechanism from windmills to regulate the engine's speed. Dickinson and Jenkins, *James Watt and the Steam Engine*, 220–221.

Page 121 The governor effectively made the steam engine into a clock. Bennett, *A History of Control Engineering, 1800–1930*; Mayr, *Authority, Liberty, and Automatic Machinery in Early Modern Europe*; Mindell, *Between Human and Machine*.

Page 121 The key breakthrough to make the engine attractive to the textile industry. Musson and Robinson, *Science and Technology in the Industrial Revolution*, 406.

Page 122 The firm gradually found successful technical managers. See Roll, *An Early Experiment in Industrial Organization* for a detailed description and analysis of Boulton and Watt business and factory practices.

Page 122 A nursery for millwrights and mechanical engineers. Dickinson and Jenkins, *James Watt and the Steam Engine*, 267, 281.

Page 122 Methodize the rotative engines. Boulton to Watt, December 7, 1782, quoted in Roll, *An Early Experiment in Industrial Organization*, 115.

Page 122 We are systematizing the business of engine making. Boulton to Smeaton, reprinted in Dickinson and Jenkins, *James Watt and the Steam Engine*, 265.

Page 123 Directions for Erecting and Working the Newly-Invented Steam Engines. Reprinted in Dickinson and Jenkins, *James Watt and the Steam Engine*, appendix II. They estimate the date as 1779, and state that only one hundred copies were printed.

Page 123 Starting a Boulton and Watt engine was a bit like playing a pipe organ. "Directions for working Rotative Engines, c. 1784," reprinted in Dickinson and Jenkins, *James Watt and the Steam Engine*, appendix III.

Page 124 An American visitor in Britain. Zachariah Allen, 1832, quoted in Hunter, *A History of Industrial Power in the United States*, 161.

Page 124 The company would receive more than a hundred orders for engines from abroad. Tann and Breckin, "The International Diffusion of the Watt Engine 1775–1825"; Kanefsky and Robey, "Steam Engines in 18th-Century Britain."

Page 125 By 1800, there were 188 reciprocating and 308 rotative engines. Schofield, *The Lunar Society of Birmingham*, 335.

Page 125 Let us be content with *doing*. Watt to Boulton, March 10, 1786, quoted in Dickinson and Jenkins, *James Watt and the Steam Engine*, 64. Emphasis in original.

CHAPTER 27

Page 127 Maintenance and repair are the most widespread forms of technical expertise. Edgerton, *The Shock of the Old*, 84; Russell and Vinsel, "Let's Get Excited about Maintenance."

Page 128 Much of what passes for innovation is actually "innovation speak." Vinsel and Russell, *The Innovation Delusion*, 13.

Page 129 "Often talk of skills." Vinsel and Russell, *The Innovation Delusion*, 118.

Page 129 A maintenance mindset is crucial to any move toward technological sustainability. Vinsel and Russell, *The Innovation Delusion*, 226.

Page 129 The Maintainers call for a positive materialism. Vinsel and Russell, *The Innovation Delusion*, 210.

CHAPTER 28

Page 132 Garry Wills found the key was Jefferson's obsession with numbers. Wills, *Inventing America*, 180.

Page 132 The "Laws of Nature," argues historian of science I. B. Cohen. Cohen, *Science and the Founding Fathers*, 122.

Page 133 Jefferson recorded in his diary not political adulation. Wills, *Inventing America*, 119.

CHAPTER 29

Page 135 A committee study for the National Research Council on government support for computer science. National Research Council, *Funding a Revolution*.

CHAPTER 30

Page 137 The infrastructure bill authorizes $1.2 trillion. For good summaries, see Badlam et al., "The US Bipartisan Infrastructure Law," "What's in the Inflation Reduction Act of 2022," and "The

CHIPS and Science Act"; Bonvillian, "Industrial Innovation Policy in the United States."

Page 138 Decarbonizing the United States may require *re*industrializing it. Meyer, "Biden's Climate Law Is Ending 40 Years of Hands-Off Government."

Page 138 Demonstration projects that prove viability. Hart, "Building Back Cleaner with Industrial Decarbonization Demonstration Projects."

Page 139 US semiconductor startups have needed to look offshore. Del Alamo et al., "Reasserting U.S. Leadership in Microelectronics."

Page 139 Much of the cost difference with US manufacturers is no longer about labor costs. Del Alamo et al., "Reasserting U.S. Leadership in Microelectronics," 17–18; Miller, *Chip War*.

Page 140 Evoked the intended response from industry. Casanova, "The CHIPS Act Has Already Sparked $200 Billion in Private Investments."

Page 140 Economists have also begun to realize how federal support shapes economic geography. Gruber and Johnson, *Jumpstarting America*.

Page 140 Manufacturing is at last being understood as the crossroads. Bonvillian, "Industrial Innovation Policy in the United States," 337.

Page 141 The new policies extend that support to later stages. Bonvillian, "Industrial Innovation Policy in the United States."

Page 141 Resistance to industrial policy. Bonvillian, *Advanced Manufacturing*, Krugman quoted at 328–329.

Page 141 Federal funds rescued Tesla from bankruptcy. Bonvillian, "Industrial Innovation Policy in the United States."

Page 142 We are at a once-in-a-century moment. Adler and Bonvillian, "America's Manufacturing Problem, and How to Fix It."

CHAPTER 31

Page 145 Wedgwood to Thomas Bentley, quoted in Smiles, *Josiah Wedgwood*, 69. The following account is synthesized and participants' quotations drawn from Hunt, *The Radical Potter*; Smiles, *Josiah Wedgwood*; and Wedgwood, *Correspondence of Josiah Wedgwood*, with additional work from Schofield, *The Lunar Society of Birmingham*, and McKendrick, "Josiah Wedgwood and Thomas Bentley," "Josiah Wedgwood and Factory Discipline."

Page 145 This combination of master and workman in one person. Quoted in Hunt, *The Radical Potter*, 42.

Page 145 Wedgwood's products cohered an English identity. Hunt, *The Radical Potter*, xvi.

Page 146 To astonish the world all at once. Quoted in Hunt, *The Radical Potter*, xix.

Page 146 Modest workshop-homes with discarded potsherds piled outside. Smiles, *Josiah Wedgwood*, 4.

Page 146 I myself began at the lowest round of the ladder. Quoted in Smiles, *Josiah Wedgwood*, 5.

Page 147 Proved the turning-point of Wedgwood's career. Smiles, *Josiah Wedgwood*, 26.

Page 147 He soon introduced a new green. Smiles, *Josiah Wedgwood*, 33.

Page 148 I saw the field was spacious. Quoted in Smiles, *Josiah Wedgwood*, 36.

Page 149 Bentley also provided the business experience. McKendrick, "Josiah Wedgwood and Thomas Bentley."

Page 150 He would sometimes copy the works by hand for later reference. Smiles, *Josiah Wedgwood*, 41.

Page 150 When prepared in perfection. Quoted in Hunt, *The Radical Potter*, 88.

Page 152 Consumers turned to ceramic pottery. Hunt, *The Radical Potter*, 28.

Page 153 Wedgwood eventually designed his own pottery lathe. To see this lathe in action, see the video: Carpentier, "Engine Turned Pottery on a 1768 Style Potters Lathe Designed by Wedgwood."

Page 154 I hope you will give this Scheme your assistance. Quoted in Schofield, *The Lunar Society*, 41.

Page 156 To place the past at the very forefront of modern, industrializing Britain. Hunt, *The Radical Potter*, 113; Keynes, "The Portland Vase."

Page 156 Through a "meticulous, iterative approach." Wedgwood, quoted in Hunt, *The Radical Potter*, 122.

Page 157 The Athenian workshops, turning out large numbers of beautiful, stylized products. Boardman, *Athenian Black Figure Vases*; Boardman, *Athenian Red Figure Vases*.

Page 157 Here we have a colony raised in a desert. Quoted in Hunt, *The Radical Potter*, 128.

Page 157 Wedgwood filled his notebooks with comments on Priestley's books. Schofield, *The Lunar Society of Birmingham*, 93.

Page 158 Last month was found under a bed of Clay. Quoted in Schofield, *The Lunar Society of Birmingham*, 96.

Page 159 Marco Polo actually coined the term *porcelain*. Hunt, *The Radical Potter*, 5.

CHAPTER 33

Page 163 I have been turning models. Quoted in McKendrick, "Josiah Wedgwood and Factory Discipline," 34.

Page 164 The "fine figure" Painters are another order. Quoted in McKendrick, "Josiah Wedgwood and Factory Discipline," 32.

Page 164 An inventory of trades at the factory. McKendrick, "Josiah Wedgwood and Factory Discipline," 33; Hunt, *The Radical Potter*, 174; McKendrick, "Josiah Wedgwood and Cost Accounting."

Page 164 A film of Wedgwood production from 1950. British Pathé, "Wedgwood Today."

Page 165 In 1790, nearly 25 percent of his workers were apprentices. McKendrick, "Josiah Wedgwood and Factory Discipline," 37.

Page 165 "My name has ben made such a scarecrow to them." Wedgwood, quoted in McKendrick, "Josiah Wedgwood and Factory Discipline," 39.

Page 166 Wedgwood found himself constantly at odds. Hunt, *The Radical Potter*, 170–172, 177.

Page 166 The utmost cleanness should be observed. Wedgwood to Bentley, July 21, 1773, in *Correspondence of Josiah Wedgwood*, 43.

Page 167 To make *Artists* . . . [of] . . . mere men. Quoted in McKendrick, "Josiah Wedgwood and Factory Discipline," 39.

Page 167 Certainly produced a team of workmen. McKendrick, "Josiah Wedgwood and Factory Discipline," 46.

Page 168 He discussed these experiments with Boulton and Priestley. Chaldecott, "Josiah Wedgwood (1730–95)—Scientist"; Wedgwood and Banks, "An Attempt to Make a Thermometer for Measuring the Higher Degrees of Heat."

Page 168 Wedgwood made dozens of his pyrometers for other researchers. Chaldecott, "Josiah Wedgwood (1730–95)—Scientist."

CHAPTER 34

Page 171 Linked to founding republican ideals. Kasson, *Civilizing the Machine*; Cochran, *Frontiers of Change*.

Page 171 So, too, with Paul Revere. The following account of Revere, and participants' quotations, are draw from from Martello, *Midnight Ride, Industrial Dawn*.

Page 176 In 1800, the US government gave Revere a $10,000 loan. Martello, *Midnight Ride, Industrial Dawn*, 228.

CHAPTER 35

Page 177 These wonderful machines, working as if they were animated beings. Quoted in Martello, *Midnight Ride, Industrial Dawn*, 217; "Hamilton's Report on Manufacturers." On other industrial developments in the early republic, see Thomson, *Structures of Change in the Mechanical Age*.

Page 178 Machines in manufacturing could be operated by women and children. Kasson, *Civilizing the Machine*, 29–30.

Page 179 A second war for independence from Britain. Cochran, *Frontiers of Change*, 76.

Page 179 In 1786, he visited England and witnessed the Boulton and Watt engine. Kasson, *Civilizing the Machine*, 22–23.

Page 179 Jefferson created a mechanized nail factory at Monticello. Kasson, *Civilizing the Machine*, 24; Gordon-Reed, *The Hemingses of Monticello*, 149, 385.

Page 179 Jefferson encouraged exploration under the commerce clause. Dupree, *Science in the Federal Government*, chapter 2; Smith, *Military Enterprise and Technological Change*, chapter 1.

CHAPTER 36

Page 181 By the time Revere retired in 1811. Martello, *Midnight Ride, Industrial Dawn*, 297.

Page 181 When the War of 1812 broke out. Martello, *Midnight Ride, Industrial Dawn*, 331.

Page 182 We are to supply some Gentlemen in New York. Revere, quoted in Martello, *Midnight Ride, Industrial Dawn*, 258.

Page 183 Watt early reached a stage. Hunter, *A History of Industrial Power*, 164–165, catalogs these critics, including Dickinson, Musson, and Robinson.

CHAPTER 38

Page 187 The Quakers promptly expelled him. Satia, *Empire of Guns*, chapter 8.

Page 189 Clarkson credited the medallions. Guyatt, "The Wedgwood Slave Medallion," 102; Bownas, *Slavery, Freedom, and Conflict*, 4–8 and 32–58; Victoria and Albert Museum, "The Wedgwood Anti-Slavery Medallion."

Page 189 A piece of propaganda central to the impassioned campaign. Hunt, *The Radical Potter*, 229; Guyatt, "The Wedgwood Slave Medallion," 94.

CHAPTER 39

Page 192 The riots continued for two days. Rose, "The Priestley Riots of 1791"; also see Schofield, *The Enlightened Priestley*, chapter 8.

Page 192 The riot seems to have had some element of class conflict. Thompson, *The Making of the English Working Class*, 73–74. Thompson's account is based on Rose, "The Priestley Riots of 1791," though his analysis places it within a longer history of class conflict. He sees the Priestley riots as disorganized, incoherent, "a late backward eddy of the transitional mob," before the populist class and worker movements of the nineteenth century.

Page 192 Watt brought a gun. Schofield, *The Lunar Society of Birmingham*, 360.

Page 193 There are few things I more regret. Priestley, quoted in Schofield, *The Lunar Society of Birmingham*, 363. On Priestley and Jefferson, see Schofield, *The Enlightened Joseph Priestley*, 342–344.

Page 193 Boulton's and Watt's sons would develop the factory into a managerial enterprise. Jones, "Living the Enlightenment and the French Revolution."

Page 193 This sum amounted to about one-quarter of the steam engines in Britain. Musson and Robinson, *Science and Technology in the Industrial Revolution*, 426; Robinson and Musson, *James Watt and the Steam Revolution*, 20.

CHAPTER 41

Page 200 Companies are learning to "stress-test" their supply chains. See, for example, Simchi-Levi, Zhu, and Loy, "Fixing the U.S. Semiconductor Supply Chain."

CHAPTER 42

Page 204 Economists are beginning to realize. Acemoglu and Johnson, *Power and Progress*.

Page 205 Generative AI will surely impact inequality. Acemoglu, Autor, and Johnson, "Can We Have Pro-Worker AI?"

Page 205 A 2024 conference at MIT on AI. Shaping the Future of Work, workshop at MIT, January 22, 2024; Coy, "A.I. Should Be a Tool, Not a Curse."

Page 206 Are often left to cobble together government support. As one example see Brewster, "This MIT Vet Wants to Revolutionize the Steel Industry."

Page 207 The United States may be entering a golden age for workers. *The Economist*, "A New Age of the Worker will Overturn Conventional Thinking."

Page 208 A post-pandemic wage compression. Autor, et. al., "The Unexpected Compression."

Page 208 Generations of underinvestment in workforce education. Adler and Bonvillian, "America's Advanced Manufacturing Problem."

Page 208 Universities have grown increasingly disconnected from industrial problems and working life. As one example, see *The Economist*, "Universities Are Failing to Boost Economic Growth."

Page 208 Familiar with the satisfaction of making things. On the moral and social satisfactions of making things, see Crawford, *Shop Class as Soul Craft*, and Sennett, *The Craftsman*.

Page 209 One approach, "augmented lean." Linder and Undheim, *Augmented Lean*, xx.

Page 210 Bringing manufacturing jobs back to the center of a city. Hatuka and Ben-Joseph, *New Industrial Urbanism*, 239.

Page 210 Yet they have been slow to adopt digital technologies. Reynolds, "Digital Technology and Supply Chain Resilience."

Page 210 Researchers are beginning to call for positive-sum automation. Armstrong and Shah, "A Smarter Strategy for Using Robots"; also see Major and Shah, *What to Expect When You're Expecting Robots*.

Page 210 Proponents of the "good jobs" movement. Ton, *The Good Jobs Strategy*; Ton, *The Case for Good Jobs*.

Page 210 A technologist career track. Liu and Bonvillian, "The Technologist."

BIBLIOGRAPHY

Accelerating Decarbonization of the U.S. Energy System. Washington, DC: National Academies Press, 2021. https://doi.org/10.17226/25932.

Acemoglu, Daron, David Autor, and Simon Johnson. "Can We Have Pro-Worker AI?" Accessed January 31, 2024. https://shapingwork.mit.edu/wp-content/uploads/2023/09/Pro-Worker-AI-Policy-Memo.pdf.

Acemoglu, Daron, and Simon Johnson. *Power and Progress: Our Thousand-Year Struggle over Technology and Prosperity.* New York: PublicAffairs, 2023.

Adler, David, and William B. Bonvillian. "America's Advanced Manufacturing Problem—and How to Fix It." *American Affairs Journal*, August 20, 2023.

Alder, Ken. *Engineering the Revolution: Arms and Enlightenment in France, 1763–1815.* Princeton, NJ: Princeton University Press, 1997.

"Alexander Hamilton's Final Version of the Report on the Subject of Manufactures [5 December 1791]." Founders Online, National Archives, https://founders.archives.gov/documents/hamilton/01-10-02-0001-0007. [Original source: Syrett, Harold C. (ed). *The Papers of Alexander Hamilton, Vol. 10, December 1791–January 1792.* New York: Columbia University Press, 1966, 230–340.]

Allen, Robert C. *The British Industrial Revolution in Global Perspective.* New Approaches to Economic and Social History. Cambridge: Cambridge University Press, 2009.

Andreessen, Marc. "The Techno-Optimist Manifesto." Andreessen Horowitz, October 16, 2023. https://a16z.com/the-techno-optimist-manifesto.

Armstrong, Ben, and Julie Shah. "A Smarter Strategy for Using Robots." *Harvard Business Review*, March 1, 2023. https://hbr.org/2023/03/a-smarter-strategy-for-using-robots.

Autor, David, Caroline Chin, Anna M. Salomons, and Bryan Seegmiller. "New Frontiers: The Origins and Content of New Work, 1940–2018." Working Paper. Working Paper Series, National Bureau of Economic Research, August 2022. https://doi.org/10.3386/w30389.

Autor, David, David A. Mindell, and Elisabeth B. Reynolds. "The Work of the Future: Building Better Jobs in an Age of Intelligent Machines." MIT Task Force on the Work of the Future, 2020. https://workofthefuture.mit.edu.

Autor, David, David A. Mindell, and Elisabeth B. Reynolds. *The Work of the Future: Building Better Jobs in an Age of Intelligent Machines*. Cambridge, MA: MIT Press, 2022.

Autor, David, Arindrajit Dube, and Annie McGrew. "The Unexpected Compression: Competition at Work in the Low Wage Labor Market." Working Paper. Working Paper Series. National Bureau of Economic Research, March 2023. https://doi.org/10.3386/w31010.

Badlam, Justin, et al. "The CHIPS and Science Act: Here's What's in It." McKinsey & Co., October 24, 2022. https://www.mckinsey.com/industries/public-sector/our-insights/the-chips-and-science-act-heres-whats-in-it.

Badlam, Justin, et al. "The US Bipartisan Infrastructure Law: Breaking It Down." McKinsey & Co., November 12, 2021. https://www.mckinsey.com/industries/public-sector/our-insights/the-us-bipartisan-infrastructure-law-breaking-it-down.

Badlam, Justin, et al. "What's in the Inflation Reduction Act (IRA) of 2022 | McKinsey." McKinsey & Co., October 24, 2022. https://www.mckinsey.com/industries/public-sector/our-insights/the-inflation-reduction-act-heres-whats-in-it.

Baynes, Ken. *The Art of the Engineer*. Woodstock, NY: Overlook Press, 1981.

Bennett, S. *A History of Control Engineering, 1800–1930*. IEE Control Engineering Series. Stevenage, NY: Peregrinus: Institution of Electrical Engineers, 1979.

Berger, Suzanne. *Making in America: From Innovation to Market*. Cambridge, MA: MIT Press, 2013.

Berger, Suzanne. "MIT Research on Production," presentation at MIT Manufacturing Symposium, May 6, 2022, slides in the author's possession.

Berry, Christopher J. *The Idea of Commercial Society in the Scottish Enlightenment*. Edinburgh: Edinburgh University Press, 2013.

Biden, Joseph R. "Press Release—The Economics of Investing in America." The American Presidency Project, July 14, 2023. https://www.presidency.ucsb.edu/documents/press-release-the-economics-investing-america.

Boardman, John. *Athenian Black Figure Vases*. New York: Oxford University Press, 1974.

Boardman, John. *Athenian Red Figure Vases: The Archaic Period: A Handbook*. World of Art. New York: Oxford University Press, 1979.

Bonvillian, William B. *Advanced Manufacturing: The New American Innovation Policies*. Cambridge, MA: MIT Press, 2017.

Bonvillian, William B. "Industrial Innovation Policy in the United States." *Annals of Science and Technology Policy* 6, no. 4: 315–411. http://dx.doi.org/10.1561/110.00000026.

Bonvillian, William B. *Technological Innovation in Legacy Sectors*. New York: Oxford University Press, 2015.

Bownas, Jane L. *Slavery, Freedom and Conflict: A Story of Two Birminghams*. Eastbourne: Sussex Academic Press, 2020.

Brewster, Lucy. "This MIT Vet Wants to Revolutionize the Steel Industry." *Fortune*, April 21, 2023. https://fortune.com/2023/04/21/an-mit-vet-wants-to-revolutionize-the-steel-industry-but-finding-funding-for-big-swings-in-climate-tech-isnt-easy.

British Pathé. "Wedgwood Today (1950)." YouTube video, 13.47. April 13, 2014. https://www.youtube.com/watch?v=tkDfmhkQyOs.

Brooks, Rodney. "Steps Toward Super Intelligence II, Beyond the Turing Test." July 28, 2018. https://rodneybrooks.com/forai-steps -toward-super-intelligence-ii-beyond-the-turing-test/.

Brynjolfsson, Erik, and Andrew McAfee. *The Second Machine Age: Work, Progress, and Prosperity in a Time of Brilliant Technologies*. New York: W. W. Norton, 2014.

Building Resilient Supply Chains, Revitalizing American Manufacturing, and Fostering Broad-Based Growth: 100-Day Reviews under Executive Order 14017. Washington, DC: The White House, June 2021. Accessed August 29, 2023. https://www.bis.doc.gov/index.php /documents/technology-evaluation/2958-100-day-supply-chain -review-report/file.

Campbell, Joseph. *The Hero with a Thousand Faces*. Princeton, NJ: Princeton University Press, 1956.

Carnegie, Andrew. *James Watt*. New York: Doubleday, Page, 1905.

Carpentier, Don. "Engine Turned Pottery on a 1768 Style Potters Lathe Designed by Wedgwood." April 6, 2014. YouTube video, 26:24. https://www.youtube.com/watch?v=n-7twF5_chU.

Casanova, Robert. "The CHIPS Act Has Already Sparked $200 Billion in Private Investments for U.S. Semiconductor Production." Semiconductor Industry Association, December 14, 2022. https:// www.semiconductors.org/the-chips-act-has-already-sparked-200 -billion-in-private-investments-for-u-s-semiconductor-production.

Chaldecott, John A. "Josiah Wedgwood (1730–95)—Scientist." *British Journal for the History of Science* 8, no. 1 (March 1975): 1–16.

Chandler, James. "Edgeworth and the Lunar Enlightenment." *Eighteenth-Century Studies* 45, no. 1 (2011): 87–104.

Charette, Robert. "The EV Transition Explained." *IEEE Spectrum*, March 31, 2023. https://spectrum.ieee.org/collections/the -ev-transition-explained.

Clagett, Martin Richard. *A Spark of Revolution (1734–1775): Thomas Jefferson, James Watt, and William Small*. Bellevue: Clyde Hill Publishing, 2021.

Cochran, Thomas C. *Frontiers of Change: Early Industrialism in America*. New York: Oxford University Press, 1981.

Cohen, I. Bernard. *Benjamin Franklin's Science*. Cambridge, MA: Harvard University Press, 1990.

Coy, Peter. "Opinion: A.I. Should Be a Tool, Not a Curse, for the Future of Work." *New York Times*, January 24, 2024. https://www.nytimes.com/2024/01/24/opinion/artificial-intelligence-ai-work-jobs.html.

Crawford, Matthew B. *Shop Class as Soulcraft: An Inquiry into the Value of Work*. Reprint edition. New York: Penguin, 2010.

Croft, Kate. "'Of Material Service to Him': Margaret Miller Watt and Ann McGregor Watt, the Wives of James Watt." In *James Watt (1736–1819)*, edited by Malcolm Dick and Caroline Archer-Parré, 61–82. *Culture, Innovation and Enlightenment*. Liverpool: Liverpool University Press, 2020. https://doi.org/10.2307/j.ctvwh8bc0.9.

Davidson, Liz. "James Watt, Slavery, and Statues." *The Hunterian Blog* (blog), August 11, 2020. http://hunterian.academicblogs.co.uk/james-watt-slavery-and-statues.

Deaton, Angus. *Economics in America: An Immigrant Economist Explores the Land of Inequality*. Princeton, NJ: Princeton University Press, 2023.

Del Alamo, Jesus, et al. "Reasserting U.S. Leadership in Microelectronics." MIT White Paper, 2021. https://hdl.handle.net/1721.1/139740.

Demidowicz, George. *The Soho Manufactory, Mint and Foundry, West Midlands: Where Boulton, Watt and Murdoch Made History*. Liverpool: Liverpool University Press, 2022.

De Vries, Jan. *The Industrious Revolution: Consumer Behavior and the Household Economy, 1650 to the Present*. Cambridge: Cambridge University Press, 2008.

Dick, Malcolm, and Caroline Archer-Parré. *James Watt (1736–1819): Culture, Innovation and Enlightenment.* Liverpool: Liverpool University Press, 2020.

Dickinson, H. W. *James Watt: Craftsman and Engineer.* Cambridge Library Collection. Technology. Cambridge University Press, 1936.

Dickinson, H. W., Rhys Jenkins, and Committee of the Watt Centenary Commemoration. *James Watt and the Steam Engine; the Memorial Volume Prepared for the Committee of the Watt Centenary Commemoration at Birmingham 1919.* Oxford: Clarendon Press, 1927.

Dupree, A. Hunter. *Science in the Federal Government, a History of Policies and Activities to 1940.* New York: Harper & Row, 1964.

Economist, The. "Economists Are Revising Their Views on Robots and Jobs." January 22, 2022. https://www.economist.com/finance-and-economics/2022/01/22/economists-are-revising-their-views-on-robots-and-jobs.

Economist, The. "A New Age of the Worker Will Overturn Conventional Thinking." November 30, 2023. https://www.economist.com/leaders/2023/11/30/a-new-age-of-the-worker-will-overturn-conventional-thinking.

Economist, The. "Universities Are Failing to Boost Economic Growth." February 5, 2024. https://www.economist.com/finance-and-economics/2024/02/05/universities-are-failing-to-boost-economic-growth.

Edgerton, David. *The Shock of the Old: Technology and Global History since 1900.* London: Profile Books, 2011.

Elm Analytics. "Ever Wonder How Many US Auto Suppliers There Really Are?" *Medium* (blog), May 19, 2017. https://medium.com/@ELMAnalytics/ever-wonder-how-many-us-auto-suppliers-there-really-are-5bee079089c8.

Energy.gov. "Adoption Readiness Levels (ARL): A Complement to TRL." Accessed January 31, 2024. https://www.energy.gov/technologytransitions/adoption-readiness-levels-arl-complement-trl.

Fordohar, Rana. "A New Technology Boom Is at Hand." *Financial Times*, March 26, 2023. https://www.ft.com/content/cb7391b2 -5d17-4806-b05c-c6125129264c.

Frey, Carl Benedikt, and Michael A. Osborne. "The Future of Employment: How Susceptible Are Jobs to Computerisation?" *Technological Forecasting and Social Change* 114 (January 2017): 254–280.

Gordon-Reed, Annette. *The Hemingses of Monticello: An American Family*. New York: W. W. Norton, 2008.

Griffith, Saul. *Electrify: An Optimist's Playbook for Our Clean Energy Future*. Cambridge, MA: MIT Press, 2021.

Gruber, Jonathan, and Simon Johnson. *Jump-Starting America: How Breakthrough Science Can Revive Economic Growth and the American Dream*. New York: PublicAffairs, 2019.

Guyatt, Mary. "The Wedgwood Slave Medallion: Values in Eighteenth-Century Design." *Journal of Design History* 13, no. 2 (2000): 93–105.

"Hamilton's Report on the Subject of Manufactures, 1791." Gilder Lehrman Institute of American History. Accessed February 1, 2024. https://www.gilderlehrman.org/history-resources/spotlight-primary -source/hamilton%E2%80%99s-report-subject-manufactures -1791.

Hatuka, Tali, and Eran Ben-Joseph. *New Industrial Urbanism: Designing Places for Production*. New York: Routledge, Taylor & Francis Group, 2022.

Hart, David M. "Building Back Cleaner with Industrial Decarbonization Demonstration Projects." Information Technology & Innovation Foundation, March 8, 2021. https://itif.org/publica tions/2021/03/08/building-back-cleaner-industrial-decarbonization -demonstration-projects.

Heilbron, J. L. *Electricity in the 17th and 18th Centuries: A Study of Early Modern Physics*. Berkeley: University of California Press, 1979.

Helper, Susan, and Evan Soltas. "Why the Pandemic Has Disrupted Supply Chains." *The White House* (blog), June 17, 2021. https://www.whitehouse.gov/cea/written-materials/2021/06/17/why-the-pandemic-has-disrupted-supply-chains/.

Horkheimer, Max, and Theodor W. Adorno. *Dialectic of Enlightenment*. New York: Continuum, 1997.

Hughes, Thomas. "In His Proper Place." *The American Interest* 3, no. 3 (Winter 2008): 117–121.

Hunt, Tristram. *The Radical Potter: The Life and Times of Josiah Wedgwood*. New York: Metropolitan Books, 2021.

Hunter, Louis C. *A History of Industrial Power in the United States, 1780–1930: Volume II Steam Power*. Charlottesville: Published for the Eleutherian Mills-Hagley Foundation by the University Press of Virginia, 1985.

Isaacson, Walter. *Benjamin Franklin: An American Life*. New York: Simon & Schuster, 2003.

Jacob, Margaret C. *Scientific Culture and the Making of the Industrial West*. New York: Oxford University Press, 1997.

Jefferson, Thomas, *Autobiography*. Accessed June 11, 2024. https://avalon.law.yale.edu/19th_century/jeffauto.asp.

Jones, Peter M. *Industrial Enlightenment: Science, Technology and Culture in Birmingham and the West Midlands 1760–1820*. Manchester: Manchester University Press, 2017.

Jones, Peter M. "Living the Enlightenment and the French Revolution: James Watt, Matthew Boulton, and Their Sons." *The Historical Journal* 42, no. 1 (1999): 157–182.

Kanefsky, John, and John Robey. "Steam Engines in 18th-Century Britain: A Quantitative Assessment." *Technology and Culture* 21, no. 2 (1980): 161–186.

Kasson, John F. *Civilizing the Machine: Technology and Republican Values in America, 1776–1900*. New York: Grossman Publishers, 1976.

Keilman, John. "America Is Back in the Factory Business." *Wall Street Journal*, April 8, 2023. https://www.wsj.com/articles/american-manufacturing-factory-jobs-comeback-3ce0c52c.

Keynes, Milo. "The Portland Vase: Sir William Hamilton, Josiah Wedgwood and the Darwins." *Notes and Records of the Royal Society of London* 52, no. 2 (1998): 237–259.

Latour, Bruno. *We Have Never Been Modern*. Cambridge, MA: Harvard University Press, 1993.

Linder, Natan, and Trond Arne Undheim. *Augmented Lean: A Human-Centric Framework for Managing Frontline Operations*. Hoboken, NJ: Wiley, 2023.

Lindqvist, Svante. "Technology on Trial: The Introduction of Steam Power Technology into Sweden, 1715–1736." PhD dissertation, Uppsala University, 1984.

Liu, John, and William Bonvillian. "The Technologist." *Issues in Science and Technology* 40, no. 2 (2024): 43–48.

Locke, Richard M. *Production in the Innovation Economy*. Cambridge, MA: MIT Press, 2014.

Major, Laura, and Julie Shah. *What to Expect When You're Expecting Robots: The Future of Human-Robot Collaboration*. New York: Basic Books, 2020.

Marsden, Ben. *Watt's Perfect Engine: Steam and the Age of Invention*. Revolutions in Science. New York: Columbia University Press, 2002.

Martello, Robert. *Midnight Ride, Industrial Dawn: Paul Revere and the Growth of American Enterprise*. Johns Hopkins Studies in the History of Technology. Baltimore, MD: Johns Hopkins University Press, 2010.

Mason, Shena, and Birmingham City Museum and Art Gallery. *Matthew Boulton: Selling What All the World Desires*. Birmingham: Birmingham City Council, 2009.

Mayr, Otto. *Authority, Liberty, and Automatic Machinery in Early Modern Europe*. Baltimore, MD: Johns Hopkins University Press, 1989.

Mazzucato, Mariana. *Mission Economy: A Moonshot Guide to Changing Capitalism*. New York: Harper Business, 2021.

McKendrick, Neil. "Josiah Wedgwood and Cost Accounting in the Industrial Revolution." *The Economic History Review* 23, no. 1 (1970): 45–67.

McKendrick, Neil. "Josiah Wedgwood and Factory Discipline." *Historical Journal* 4, no. 1 (1961): 30–55.

McKendrick, Neil. "Josiah Wedgwood and Thomas Bentley: An Inventor-Entrepreneur Partnership in the Industrial Revolution." *Transactions of the Royal Historical Society* 14 (December 1964): 1–33.

Meyer, Robinson. "Biden's Climate Law Is Ending 40 Years of Hands-Off Government." *The Atlantic*, August 18, 2022.

Miller, Chris. *Chip War: The Fight for the World's Most Critical Technology*. New York: Scribner, 2022.

Mindell, David A. *Between Human and Machine: Feedback, Control, and Computing before Cybernetics*. Johns Hopkins Studies in the History of Technology. Baltimore, MD: Johns Hopkins University Press, 2002.

Mindell, David A. *Digital Apollo: Human and Machine in Spaceflight*. Cambridge, MA: MIT Press, 2011.

Mindell, David A. *Our Robots, Ourselves: Robotics and the Myths of Autonomy*. New York: Viking, 2015.

Mokyr, Joel. *A Culture of Growth: The Origins of the Modern Economy*. The Graz Schumpeter Lectures. Princeton, NJ: Princeton University Press, 2016.

Mokyr, Joel. *The Enlightened Economy: An Economic History of Britain, 1700–1850*. The New Economic History of Britain. New Haven, CT: Yale University Press, 2009.

Mokyr, Joel. *The Gifts of Athena: Historical Origins of the Knowledge Economy*. Course Book. Princeton, NJ: Princeton University Press, 2011.

Mullen, Stephen. *The Glasgow Sugar Aristocracy: Scotland and Caribbean Slavery, 1775–1838*. New Historical Perspectives. London: University of London Press, 2022.

Musson, A. E. (Albert Edward), and Eric Robinson. *Science and Technology in the Industrial Revolution*. Manchester: Manchester University Press, 1969.

National Research Council. *Funding a Revolution: Government Support for Computing Research*. Washington, DC: National Academies Press, 1999.

Padhi, Asutosh. *The Titanium Economy: How Industrial Technology Can Create a Better, Faster, Stronger America*. New York: PublicAffairs, 2022.

Peaucelle, Jean-Louis, and Cameron Guthrie. "How Adam Smith Found Inspiration in French Texts on Pin Making in the Eighteenth Century." *History of Economic Ideas* 19, no. 3 (2011): 41–68.

Porter, Roy. *Enlightenment: Britain and the Creation of the Modern World*. London: Allen Lane/Penguin Press, 2000.

Postelwait, Jeff. "Transformative Times: Update on the U.S. Transformer Supply Chain." T&D World, July 12, 2022. https://www .tdworld.com/utility-business/article/21243198/transformative -times-update-on-the-us-transformer-supply-chain.

Reynolds, Elisabeth. "Digital Technology and Supply Chain Resilience: A Call to Action to Accelerate U.S. Manufacturing Competitiveness." Massachusetts Business Roundtable, Manufacturing@ MIT, December 2023.

Reynolds, Elisabeth B., Hiram M. Samel, and Joyce Lawrence. "Learning by Building: Complementary Assets and the Migration of Capabilities in U.S. Innovative Firms." In *Production in the Innovation Economy*, edited by Richard Locke and Rachel Wellhausen. Cambridge, MA: MIT Press, 2014.

Riskin, Jessica. *Science in the Age of Sensibility: The Sentimental Empiricists of the French Enlightenment*. Chicago: University of Chicago Press, 2002.

Riskin, Jessica. *The Restless Clock: A History of the Centuries-Long Argument over What Makes Living Things Tick*. Chicago: University of Chicago Press, 2016.

Robertson, Ritchie. *The Enlightenment: The Pursuit of Happiness, 1680–1790*. New York: Harper, 2021.

Robinson, Eric. *Partners in Science: Letters of James Watt and Joseph Black*. Cambridge, MA: Harvard University Press, 1970.

Robinson, Eric, and Musson, A. E. *James Watt and the Steam Revolution: A Documentary History*. London: Adams & Dart, 1969.

Roll, Eric. *An Early Experiment in Industrial Organization; Being a History of the Firm of Boulton & Watt, 1775–1805*. New York: A. M. Kelley, 1968.

Rose, R. B. "The Priestley Riots of 1791." *Past & Present*, no. 18 (1960): 68–88.

Rosen, William. *The Most Powerful Idea in the World: A Story of Steam, Industry, and Invention*. New York: Random House, 2010.

Russell, Andrew, and Lee Vinsel. "Opinion: Let's Get Excited About Maintenance!" *New York Times*, July 22, 2017.

Russell, Ben. *James Watt: Making the World Anew*. London: Reaktion Books, 2014.

Satia, Priya. *Empire of Guns: The Violent Making of the Industrial Revolution*. Stanford, CA: Stanford University Press, 2018.

Schofield, Robert E. *The Enlightenment of Joseph Priestley: A Study of His Life and Work from 1733 to 1773*. University Park: Pennsylvania State University Press, 1997.

Schofield, Robert E. "The Industrial Orientation of Science in the Lunar Society of Birmingham." *Isis* 48, no. 4 (1957): 408–415.

Schofield, Robert E. *The Lunar Society of Birmingham: A Social History of Provincial Science and Industry in Eighteenth-Century England*. Oxford: Clarendon Press, 1963.

Schwab, Klaus. *Fourth Industrial Revolution*. London: Penguin Group, 2017.

Sennett, Richard. *The Craftsman*. New Haven: Yale University Press, 2008.

Shapin, Steven. *The Scientific Revolution*. Chicago: University of Chicago Press, 2018.

Sheffi, Yosef. *The Magic Conveyor Belt: Supply Chains, A.I., and the Future of Work*. Cambridge, MA: MIT CTL Media, 2023.

Simchi-Levi, David, Feng Zhu, and Matthew Loy. "Fixing the U.S. Semiconductor Supply Chain." *Harvard Business Review*, October 25, 2022.

Smil, Vaclav. *How the World Really Works: The Science Behind How We Got Here and Where We're Going*. New York: Viking, 2022.

Smil, Vaclav. *Made in the USA: The Rise and Retreat of American Manufacturing*. Cambridge, MA: MIT Press, 2013.

Smil, Vaclav. *Power Density: A Key to Understanding Energy Sources and Uses*. Cambridge, MA: MIT Press, 2015.

Smiles, Samuel. *Josiah Wedgwood, His Personal History*. London: Routledge/Thoemmes Press, 1997.

Smiles, Samuel. *Lives of Boulton and Watt*. Collected Works of Samuel Smiles. London: Routledge/Theommes, 1997.

Smiles, Samuel. *Self Help: with illustrations of character, conduct and perseverance*. London: J Murray, 1859.

Smith, Adam. *An Inquiry into the Nature and Causes of the Wealth of Nations*. New York: Modern Library, 1994.

Smith, Merritt Roe. *Military Enterprise and Technological Change: Perspectives on the American Experience*. Cambridge, MA: MIT Press, 1985.

Stone, Brad. *The Everything Store: Jeff Bezos and the Age of Amazon*. New York: Little, Brown, 2013.

Tann, Jennifer, and M. J. Breckin. "The International Diffusion of the Watt Engine, 1775–1825." *Economic History Review* 31, no. 4 (1978): 541–564.

Tatar, Maria. *The Heroine with 1001 Faces*. New York: Liveright Publishing Corporation, 2021.

Thompson, E. P. "Time, Work-Discipline, and Industrial Capitalism." *Past & Present*, no. 38 (1967): 56–97.

Thompson, E. P. (Edward Palmer). *The Making of the English Working Class*. New York: Pantheon Books, 1964.

Thomson, Ross. *Structures of Change in the Mechanical Age: Technological Innovation in the United States, 1790–1865.* Johns Hopkins Studies in the History of Technology. Baltimore, MD: Johns Hopkins University Press, 2009.

Tirpak, John. "New Defense Industrial Base Strategy Warns of Long Recovery to Reverse Atrophy." *Air & Space Forces Magazine*, January 12, 2024. https://www.airandspaceforces.com/new-defense -industrial-base-strategy-long-recovery.

Ton, Zeynep. *The Case for Good Jobs: How Great Companies Bring Dignity, Pay and Meaning to Everyone's Work.* Boston: Harvard Business Review Press, 2023.

Ton, Zeynep. *The Good Jobs Strategy: How the Smartest Companies Invest in Employees to Lower Costs and Boost Profits.* Boston: New Harvest/Houghton Mifflin Harcourt, 2014.

Uglow, Jennifer S. *The Lunar Men: Five Friends Whose Curiosity Changed the World.* New York: Farrar, Straus and Giroux, 2002.

United States Department of Defense. "National Defense Industrial Strategy." 2023.

US Department of Commerce, Bureau of Industry and Security, Office of Technology Evaluation. "The Effect of Imports of Transformers and Transformer Components on the National Security." October 15, 2020.

Victoria and Albert Museum. "The Wedgwood Anti-Slavery Medallion V&A." Accessed February 1, 2024. https://www.vam.ac.uk /articles/the-wedgwood-anti-slavery-medallion.

Vinsel, Lee, and Andrew L Russell. *The Innovation Delusion.* New York: Crown Currency, 2020.

Voskuhl, Adelheid. *Androids in the Enlightenment: Mechanics, Artisans, and Cultures of the Self.* Chicago: University of Chicago Press, 2013.

"'We Have Met the Enemy and He Is Us.'" Billy Ireland Cartoon Library & Museum, Ohio State University, January 5, 2020. https:// library.osu.edu/site/40stories/2020/01/05/we-have-met-the-enemy.

Weber, Max. *The Protestant Ethic and the "Spirit" of Capitalism*. New York: Charles Scriber's Sons, 1958.

Wedgwood, Josiah. *An Address to the Young Inhabitants of the Pottery, by Josiah Wedgwood, F.R.S. Potter to Her Majesty*. Newcastle: J. Smith, 1783.

Wedgwood, Josiah, and Joseph Banks. "An Attempt to Make a Thermometer for Measuring the Higher Degrees of Heat, from a Red Heat up to the Strongest That Vessels Made of Clay Can Support." *Philosophical Transactions of the Royal Society of London* 72 (January 1782): 305–326.

Wedgwood, Josiah. *Correspondence of Josiah Wedgwood*. Edited by Katherine Euphemia Farrer. Cambridge Library Collection. Cambridge: Cambridge University Press, 2010.

Wiencek, Henry. "The Dark Side of Thomas Jefferson." *Smithsonian Magazine*, October 2012. https://www.smithsonianmag.com/history/the-dark-side-of-thomas-jefferson-35976004.

Williams, Raymond. *Keywords: A Vocabulary of Culture and Society*. New edition. Oxford New York: Oxford University Press, 2014.

Williams, Rosalind H. *Dream Worlds: Mass Consumption in Late Nineteenth Century France*. Berkeley: University of California Press, 1982.

Wills, Garry. *Inventing America: Jefferson's Declaration of Independence*. Garden City, NY: Doubleday, 1978.

Wood, Gordon S. *The Americanization of Benjamin Franklin*. New York: Penguin Press, 2004.

Wood, Gordon S. *The Radicalism of the American Revolution*. New York: Vintage Books, 1993.

INDEX

Additive manufacturing (3D printing), 6, 75, 197, 199
Advanced Manufacturing Partnership, 74
AFL-CIO, 209
Aircraft, and maintenance, 128
Air pump, 84
Albion Mill, 125, 179
Amazon, 59, 74
American Enlightenment, 172
American Revolution, 16, 53–54, 188
    and the Lunar Society, 107
    and Paul Revere, 171–172, 173, 177
Ammonia, 7, 138
Anderson, John, 83, 84
Anti-saccharite movement, 187
Apollo lunar lander, 92
Apple, 27–29, 139
Artificial intelligence (AI), 6, 75, 141, 197
    generative, 169
Artisans, rise to prominence, 65
Athenian Greeks, 156–157
Atlantic, The (magazine), 138
Augmented lean manufacturing, 209

Austen, Jane, 110, 145–146
Automation, 78, 91–93
Automobiles, and maintenance, 127–128
Autonomous cars, 75, 102
Autonomy, 6
Autor, David, 75, 76, 78
Aviation, progress in, 161–162, 185

Babbage, Charles, 32, 41
Bacon, Francis, 61
Banks, Joseph, 45
Baptists, 52
Barriers, to new technologies, 136
Baskerville, John, 43
Bastille day dinner riots (Birmingham), 191–193
Bell making, 174
Bell Works, 148
Bentley, Thomas, 149
Berger, Suzanne, 34, 74
B-52 bomber, 128
Biden administration, 56, 137
"Big four" commodities, 7
Bipartisan Infrastructure Law, 137

Birmingham, England, 23–25,
    69
  Bastille day dinner riots,
    191–193
  Franklin's visit to, 43–44
Black, Joseph, 83–84, 85, 88,
    117, 168
Blake, William, 125
Blockchain, 135
Bloomfield Colliery, 99
Boeing, 203
Boilers, steam, 182
Bonvillian, William, 33, 35
Boston Mechanic Organization,
    174
Boswell, James, 62, 99
Botanic Garden (Erasmus
    Darwin), 189
Boulton, Ann, 70
Boulton, Matthew, 39–40, 153,
    182–183, 218
  in Cornwall, 117
  and Erasmus Darwin, 39–42
  at first recorded meeting of
    the Lunar Society, 106
  framed relationship with
    Watt with love, 98
  and Franklin, 43–44
  ideas for a steam engine
    factory, 97–98
  interest in electricity, 41–42
  introduced to Watt, 71
  and the Lunar Circle, 45–46
  Roebuck's offer, 96
  and Small, 63–64, 71
  Smiles's biography of, 67
  Soho manufactury, 70–71
    vision of steam engine in
      industry, 96–97
    as Watt's business partner, 3–
      4, 117–118, 213
Boyle, Robert, 43, 84
Bridgewater, Duke of, 154
Bridgewater Canal, 52–53
Brindley, James, 107, 154, 157
British Library, 23–24
British Navy, 23
Brooklyn Navy Yard, 101
Brooks, Rodney, 77
Brunel, Isambard Kingdom, 66
Buckingham Palace, 24
Bull, John, 183
Burke, Edmund, 191
Burslem, England, 146–147,
    155
Bush, Vannevar, 28

California, and Apple, 27–29
Campbell, Joseph, 67–68
Canton, Massachusetts, 176, 181
Carnegie, Andrew, 1, 196
Catherine the Great, of Russia,
    152
CES trade show, 209
Change, resistance to, 136
Charlotte, Queen of England,
    24, 45, 151, 205
China, 38
  and Apple, 27
  and chips, 139
  and process innovation, 29,
    35
CHIPS and Science Act, 137,
    138–140

Clarkson, Thomas, 188–189
Clermont (steamship), 182
Climate change, 5, 38, 138
Clock metaphor, 84
"Club of Potters," Wedgwood's
    proposal for, 106
Coal and coal mines, 71, 115–
    116
Cognitive science, 75
Cohen, I. B., 132
Collaboration, 12
College of William and Mary,
    62–63
Commonplace books, Lunar
    Society's members', 111
Concrete, 7, 138
Constitution, USS, 175, 178
Copley, John Singleton, 171–
    172
Copper cladding, 175, 181
Cornwall, 116–117
COVID-19 pandemic, 49, 55,
    137, 140
Coxe, Tench, 177–178
Craftsmen and craftsmanship.
    See individual craftsmen, e.g.
    Revere, Paul
Craft-type jobs, deskilling, 76,
    169
Cuba, 127–128
Cults of newness, 135–136
Culture, 9, 12

Darwin, Erasmus, 40–41
    and Boulton, 40, 41–42
    at first recorded meeting of
        the Lunar Society, 106

and Franklin, 43–44
    interest in canals, 154–155
    interest in electricity, 41–42
    and the Lunar Circle, 45–46
    and Priestley, 111
    and the Wedgwoods, 69–70
Day, Thomas, 107
DC-3 airliner, 161–162
Deaton, Angus, 33
Decarbonization, 38, 138
Declaration of Independence,
    131–133
Design, and manufacturing,
    27–29
Deskilling, 76, 169
Diderot, Denis, 32, 62
Digital computers and
    computing, 91–92
"Directions for Erecting and
    Working the Newly
    Invented Steam Engines.
    By Boulton and Watt." 123
*Directions for Impregnating Water
    with Fixed Air* (Priestley),
    108
Disruption and disruptions, 6,
    49, 59, 102
Dissenters, 52
Dissenting academies, 52–53
Division of labor, 32–33
    and Boulton, 39–40
    in Wedgwood's Etruria
        factory, 164–168
DIY renaissance, 208
Domesticity and professional
    life, 109
Douglas Aircraft, 161–162

Dr. Seuss, 207
Dr. X, 110
Duopolies, 200
Dynamism, 59, 207

East India Company, 152
Economics and economists,
    33–34
  industry agnosticism, 34
Edgerton, David, 127–128
Edgeworth, Maria, 110, 218
Edgeworth, Richard, 110
Edison, Thomas Alva, 161
Eiffel Tower, 101
Electricity, core vocabulary of,
    42–43
Electric vehicles (EVs), 48, 141
Electrification, 6, 38
Encyclopaedia Britannica, 61–62
Encyclopédie, ou Dictionnaire
    Raisonné des Sciences, des Arts
    et Des Métiers (Diderot),
    32, 62
Engels, Friedrich, 166
Engine, meanings of, 33
Engineer-heroes, 68
Engineers, rise to prominence,
    65
Enlightenment, 15–17,
    20. *See also* Industrial
    Enlightenment
  and Jefferson, 131–133
  Scottish, 61–63, 132–133
Erie Canal, 185
Essential, workers, 55–56
Etruria factory, 155–159
  film of, 164–165

organization of, 163–168
  and the pyrometer, 168–169
Etruscans, 155–157
Evans, Oliver, 182, 183

Faraday, Michael, 52
Final assembly line, 47
First day vases, 163, 213
Flexible work, 6
Flyball governor, 2, 120–121
Ford, Henry, 47
Ford Motor Corporation, 75
  assembly plants, 47
  and supply chain, 56
Fossil fuels, 7
Fothergill, John, 157
Foucault, Michel, 16
Fourth Industrial Revolution,
    11
Franklin, Benjamin, 108, 171
  autobiography, 66–67
  and electricity, 42–43
  and Enlightenment values,
    53–54
  and the Lunar Circle, 45–46
  and Priestley, 111
  and Smiles, 66–67
  visit to Birmingham, 43–44
French Revolution, 16, 188,
    191
"Frog Service" (Wedgwood
    pottery), 152
Fulton, Robert, 182

Galton, Samuel, Jr., 187
General Electric, 101
Generative AI, 169

George III, King of England, 23–24, 45, 125, 151, 168, 192
Germany
  competition with US manufacturing, 73
  Fraunhofer Institutes, 74
Gibbon, Edward, 132, 145
GitHub, 198–199
Glasgow, Scotland, 81, 83
GlobalFoundries, 140
Grand Trunk Canal, 155
Graphical language, Watt's need for, 87
Growth, innovation first model of, 34

Hackwood, William, 164, 188
Hamilton (musical), 177
Hamilton, Alexander, 177–178, 179
Hamilton, Lady, 45
Hamilton, Lord William, 45, 155–157
Hard tech, 211–212
Harper's Ferry, Virginia (arsenal), 179
Harrison, John, 81
Hephaestus, 67
Herculaneum, 111, 155
Heroic mythology of inventors, Smiles's, 66–68
Hero of a Thousand Faces, The (Campbell), 67–68
Herschel, William, 45
History and Present State of Electricity (Priestley), 52, 108, 111

History of the Decline and Fall of the Roman Empire, The (Gibbon), 131, 145
Horsepower, 2, 116
Humanism and the humanities, 13
Human-robotic collaboration, 6
Hume, David, 62
Humphreys, Joshua, 176
Hunt, Tristram, 156, 189
Hunter, Louis, 183
Hutton, James, 62, 107
Hutton, William, 24, 25, 51, 192

IBM, 75
Individual action, 12
Industrial Enlightenment, 19–21, 38, 133
  and the early United States, 171
Industrial enlightenment, new, 9–10, 197–201, 217–218. *See also* New industrialism
Industrial improvement, human dimensions of, 12
Industrial Revolution, 15–17
  and Birmingham, 24–25
  and Franklin, 53
  and Jefferson, 131–133
Industrial revolutions, 11–13
Industrious Revolution, 51–52, 65–66, 152, 172
Industry, 48–49
  as a human virtue, 51
  and new industrialism, 212

Industry, transformation of,
        5–6, 197–201
    and industrial systems, 7–9,
        49, 57, 136
    Industry 4.0, 9, 11
    and federal policy, 140–143
Industry 4.0, 9, 11, 135
Inflation Reduction Act (IRA),
        137–138
Innovation, process and
        product, 29, 185–186
Intel, 139, 140
iPhone, 29, 161
Isaacson, Walter, 54

Japan, competition with US
        manufacturing, 73
Jasperware, 159, 164
Jefferson, Thomas, 125, 138,
        171, 176, 213
    "lost world" of, 131–133
    and Revere, 176–177, 181
Jobs, 210–211
    and autonomy of workers, 92
    and shift in the economy,
        77–78
Jobs, Steve, 27
Job training resources, 142
Johnson, Samuel, 99
Jung, Carl, 67

Keir, James, 40, 62, 106, 111,
        191
Kinneil engine, 114
Kinneil House, 89, 90
Knowledge, producing: and
        producing things, 28
Krugman, Paul, 33

Labor, division of, 32–33
    and Boulton, 39–40
    in Wedgwood's Etruria
        factory, 164–168
Labor unions, 209
Lap engine, valve gear on, 86
Latent heat, 83–84, 85
Lavoisier, Antoine, 109, 168
Leaders for Global Operations
        (MIT), 73–74
Leaders for Manufacturing
        program (MIT), 73–74
Lead poisoning, 166–167
Lean manufacturing, 209
Led Zeppelin, 24
Legacy sectors, of US economy,
        34
Lewis and Clark Expedition,
        179–180
Lightning rod, Franklin's 43
Lives of the Engineers (Smiles), 66
Livingston, Robert, 182
Locke, John, 132
Lockheed Martin, 203
Locomotives, steam, 183
Longfellow, Henry Wadsworth,
        171
Lucas, George, 67
Lunar Circle, 45–46, 61, 64, 69.
        See also Lunar Society
        Watt joins, 89–90
Lunar Society, 13, 106–111,
        168. See also Lunar Circle;
        New industrialism
    and the Bastille day riots,
        191–193
    and the Enlightenment,
        15–17

optimism of, 213–214
passing of members, 194
and social change, 189
Lunar Society, new, 197

MacGregor, Ann, 113
Machinery, and the Scientific
    Revolution, 84
Made in America (MIT), 73–74
Maintainers, the, 129, 203
Maintenance, 127–129
Makers, 203
Making in America (MIT), 74
Malthus, Thomas Robert, 149
Manufacturing, 34–35, 37–38
    and design, 27–29
    history of, 19
    and industry, 48
    knowledge, 198
    MIT studies of problems in,
        73–79
    and software, 198
Manufacturing USA institutes,
    74
Market equilibria, 33–34
Martello, Robert, 173–174
Marx, Karl, 16
Material flows, 47
Mechanics, rise to prominence,
    65
Medallion, anti-slavery, 188–
    189
Metallurgy, 39–40
Metaverse, the, 135
Metropolitan Transit Authority
    (MTA), 101–102
Meyer, Robinson, 138
Micron, 140

Middle economy, 38
Millwrights, 115
MIT, 12, 205, 212
    Leaders for Manufacturing
        program, 73–74
    studies of problems in, 73–79
Mobility, transformation in,
    199–200
Monopolies, 200
Moore's law, 29
Murdock, William, 118, 120,
    122, 182
Museum of Fine Arts (MFA), in
    Boston, 171, 172

National defense, as industrial
    policy, 180
National Research Council, 135
Naval Act of 1794, 178–179
Nelson, Lord, 45
Newcomen, Thomas, 1, 52
Newcomen engines, 1, 70–71,
    84, 88
New industrialism, 203
    human-centeredness of,
        207–211
    and human dimensions of
        technology, 204–206
    and national culture, 204
    and public recognition of
        dependence on industrial
        systems, 203–204
    and technology, 204–206,
        211–212
    and venture capital, 206–207
Newton, Isaac, 15, 61, 65
    mechanical philosophy, 84
    Principia, 132–133

New York City subway, 101–
103, 136
Noncomformists, 52
Norden bombsight, 91
Northrop Grumman, 203

Obama administration, 74,
141
Odyssey (Homer), 68
Operation Warp Speed, 140
Organizational learning, 37
Ormolu, 40
Oxford University, 52–53

Parliament, 183, 189
Pearl Street Station, 161
PepsiCo, 75
*Philosophical Transactions of the
Royal Society*, 41
Phlogiston and de-
phlogisticated air, 109
Pin factories, 32–33
PitchBook, 206
Plant, Robert, 24
Plastics, 7, 138
Polo, Marco, 159
Pompeii, 110–111, 155
Poor Richard's Almanack
(Franklin), 172
Porcelain, 159
Porter, Roy, 16
Ports, and essential workers, 56
Potters, 164–166
"Potters' Instructions" (Etruria
factory), 165
Pottery, 150. *See also*
Wedgwood, Josiah

Practical Education (Edgeworth
and Edgeworth), 110
Priestly, Joseph, 43, 52–53,
107–109
and the Lunar Society, 108–
109, 111
Principia (Newton), 132–133
Process innovation, 4, 29, 185–
186
Product innovation, 4, 29,
185–186
Productivity, and workers, 77
Protestant ethic, 53
Pugh, Francis, 87
Puritans, 52
Pyrometer, 168–169

Quakers, 52, 187
Qualcomm, 140
"Queen's Ware" (Wedgwood
pottery), 151

Radar, 91
Rainhill Trials, 183
Remote work, 6
Repair Rather than Replace,
advocates of, 203
"Report on Manufacturers"
(Hamilton), 177
Revere, Paul, 171–176, 181–
182
Revolutionary War, American,
16, 53–54, 188
and the Lunar Society, 107
and Paul Revere, 171–172,
173, 177
Reynolds, Elisabeth, 75

Reynolds, Maria, 177

Right to Repair, advocates of, 203

Robinson, Anne, 44

Robinson, Mary, 40

Robison, John, 81, 84, 85

Robots and robotics, 6, 74–75, 197, 198–199

Rocket (steam locomotive), 183

Roebuck, John, 88, 89, 95, 96–97, 98

Rotary (rotative) engine, 122–123, 124–125

Rotary motion, 119–120

Rousseau, Jean-Jacques, 149

"Royal Pattern" (Wedgwood pottery), 151

Royal Society of London, 84, 168
  and Franklin, 43

Russel, Andrew, 128–129

Russia, and Ukraine, 56

Scaling up, 88

Schofield, Robert, 46, 99, 106, 107–108

Schwab, Klaus, 11

Schweppe, Johann, 108

Science the Endless Frontier (Bush), 28

Scientific instruments, and Watt's workshop (Glasgow), 82–83

Scientific Revolution, The, 15, 84

Scottish Enlightenment, 61–63, 132–133

Self help, 65–66

Self Help (Smiles), 66

Semiconductor chips, 56, 139

Seven Years' War, 23

Shapin, Steven, 84

"Shaping the Future of Work" (MIT conference on AI), 205

Sheffi, Yossi, 57

Skills, 129
  and deskilling, 76

Slavery, 107, 187–189

Small, William, 61–64, 82, 89, 96, 155, 180
  and Boulton, 63–64
  death, 61, 105–106
  introduces Boulton to Watt, 71
  and Jefferson, 132, 213

Smil, Vaclav, 7

Smiles, Samuel, 65–67, 68, 90

Smith, Adam, 31–33, 39–40, 44–45, 62, 81, 131, 141, 213

Social change, and how we make things (manufacturing), 38, 189

Society for Establishing Useful Manufactures, 178

Society for the Abolition of the Slave Trade, 188

Society for the encouragement of Arts and Manufacturing, 62–63

Software
  and industrial transformation, 200–201
  and manufacturing, 198

SoHo (Birmingham), 45, 69, 70
Soho Foundry, 98, 110, 123
Solyndra, 141
"Sons of Liberty Bowl"
    (Revere), 171–172
Soviet spacecraft, 92
Springfield Armory, 177, 179
Startups, 28–29
Star Wars (film), 67
Steam engine. *See also* Watt,
    James
  Boulton's ideas for a steam
    engine factory, 97–98
  complexities introduced by
    separate condenser, 86–87
  Kinneil engine, 114
  noncondensing, high
    pressure, 182–183
  operating, 123–124
  patents, fees, and royalties,
    116, 118–119, 121–122
  piston seals, 90
  precision cylinder, 115
  rotary (rotative), 122–123,
    124–125
  and rotary motion, 119–120
  separate condenser, 86, 87
  and Wilkinson, 114–115
Steel, 7, 138
Stephenson, George, 67
Stephenson, John, 183
Stephenson, Robert, 67, 183
Stevens, John, 183
Stoddert, Benjamin, 175–176
Subway, New York City, 101–
    113

Supply and demand, 33–34
Supply chains, 37, 55–56,
    57–59
  breakdowns of, 5
  as networks, 57

Taiwan, 56
Taiwan Semiconductor
    Manufacturing
    Corporation (TSMC), 29,
    139, 140
Tea
  and the Industrious
    Revolution, 51–52
  and Wedgwood pottery,
    151–152
Technology
  human dimensions of, 204–
    206
  US government influence on,
    185–186
Techno optimism, 9
Tesla, 141
Thoreau, Henry David, 138
3D printing. *See* Additive
    manufacturing
Tipton Chemical Works, 111
Toilet paper crisis, 55–56
Tolkien, J. R. R., 24
Townshend Acts of 1767, 172
Treaty of Paris (1763), 23
Trent and Mersey Canal, 69,
    154
Trevithick, Richard, 183
Tulip, 209
Turner, Matthew, 148–149

Uglow, Jenny, 17, 109
Ukraine, 56, 203
Unitarians, 52, 188
United States
    Air Force, 128
    Army engineers, 185
    Congress, 177, 179
    Constitution, 179–180
    Census Bureau, 77
    and the challenge of
        adoption, 135–136
    Department of Defense, 7,
        204
    federal influence on
        technology, 185–186
    Navy, 175, 178–179
    research and development
        (R&D) funding, 140, 186
    transformation of policy
        towards industry, 140–
        143
University of Glasgow, 81,
    82
Urban technology, 75
US Defense Production Act,
    204
Utopian visions, 20–21

Venture capital, and new
    industrialism, 206–207
Vietnam, 27
Vinsel, Lee, 128–129
Virtual presence, 6
Virtual reality (VR), 135
Visionaries, 135
Volta, Alessandro, 52

War of 1812, 181
Warrington Academy, 52–53,
    108, 149, 187
Washington, George, 63, 177,
    178–179
Watt, Ann, 110, 118, 218
Watt, James, 1–4, 82–84, 182–
    183, 218
    and the Bastille day riots,
        192–193
    and Boulton as business
        partner, 3–4, 96, 117–118,
        213
    Boulton's ideas for a steam
        engine factory, 97–98
    and Boulton's larger vision
        for steam in industry, 95,
        96–97
    complexities introduced by
        separate condenser, 86–
        87
    eureka moment, 85
    at first recorded meeting of
        the Lunar Society, 106
    funds for experimental
        engine work, 88–90
    Glasgow workshop, 81
    gloominess, 40, 90, 105
    introduced to Boulton, 71
    inventions, 118
    and mechanical drawing, 87
    model of steam engine,
        85–86
    move to Cornwall, 116–117
    need for a mechanical
        industry, 87

Watt, James (cont.)
  and patents, fees, and
    royalties, 96, 116, 118–
    119, 121–122
  Robinson points attention to
    Newcomen engines, 81
  and Roebuck, 88–89
  Smiles's biography of, 67
  withdrawal from steam
    engine business, 193–194
  wives of, 109–110, 113
Watt, James, Jr., 110
Watt, Margaret, 110
Watt, Peggy, 83, 98, 109–110,
  113
Wealth of Nations, The (Smith),
  31–32, 110, 131, 159
Wealth work, 78
Weber, Max, 53
Web3, 136
Wedgwood, Emma, 70
Wedgwood, Josiah, 145–146,
  218
  anti-slavery of, 187–189
  and Bentley, 65, 149–151
  and Boulton, 153
  concerns about distribution,
    153–155
  and the Darwins, 69–70
  early years, 146–148
  Etruria factory (*see* Etruria
    factory)
  first business, 148–149
  first day vases, 163, 213
  global customers, 150–153
  and the Lunar Circle, 106,
    154

  and the Lunar Society, 157–
    159
  and Priestley, 111
  Smiles's biography of, 67
  and Turner, 148–149
Wedgwood, Susannah, 69–70
Wedgwood black basalt, 163
West Point, U. S. Military
  Academy at, 180
Whieldon, Thomas, 147
Whirling governor, 120–121
Whitehurst, John, 41, 45–46,
  105, 158–159
Wilke, Jeff, 74, 75
Wilkinson, John, 107, 108,
  114–115, 123, 205
Wilkinson, Mary, 108
Wilkinson boring machine,
  116
Williamsburg, Virginia, 62–63
Wills, Gary, 132
Withering, William, 106, 192
Wood, Gordon, 46, 53
Woods Hole Oceanographic
  Institution, 12
Work, 12
  anthropology of, 75
  future of, 75
  and technology in the
    workplace, 92
  transformations of, 6
Workers
  essential, 55–56
  in the Etruria factory, 164–
    168
  and productivity, 77
  and skills, 129

in Soho, 122
STEM, 138
"Work of the Future, The"
    (MIT), 76, 78–79
World War II
  and automation, 91
  and aviation, 185
Wright, Joseph, 107
Wright brothers, 161, 185

YouTube, 208

Zimmerman, Trevor, 207